Guide to Purchasing Electricity and Gas

By Paul R. Cunningham, P.E.
and
David Burrell, C.E.M.

Guide to Purchasing Electricity and Gas

By Paul R. Cunningham, P.E.
and
David Burrell, C.E.M.

Published by
THE FAIRMONT PRESS, INC.
700 Indian Trail
Lilburn, GA 30047

Library of Congress Cataloging-in-Publication Data

Cunningham, Paul R., 1930-
Guide to purchasing electricity and gas / by Paul R. Cunningham and
David Burrell.
    p. cm.
    Includes bibliographical references (p.   ) and index.
    ISBN 0-88173-293-1
    1. Electric power--United States--Purchasing. 2. Electric power--Canada--
    Purchasing. 3. Natural gas--United States--Purchasing. 4. Natural gas--
    Canada--Purchasing. 5. Electric utilities--North America. 6. Gas industry
    --North America. I. Burrell, David, 1940-. II. Title.
HD9685.U5C86   1999   333.793'2'0687--dc21   98-38594
                                                                CIP

*Guide to purchasing electricity and gas / by Paul R. Cunningham and
David Burrell.*

Published by The Fairmont Press, Inc.
700 Indian Trail
Lilburn, GA 30047

Printed in the United States of America

10 9 8 7 6 5 4 3 2 1

ISBN 0-88173-293-1 FP

ISBN 0-13-012652-7 PH

While every effort is made to provide dependable information, the publisher, authors, and
editors cannot be held responsible for any errors or omissions.

Distributed by Prentice Hall PTR
Prentice-Hall, Inc.
A Simon & Schuster Company
Upper Saddle River, NJ 07458

Prentice-Hall International (UK) Limited, London
Prentice-Hall of Australia Pty. Limited, Sydney
Prentice-Hall Canada Inc., Toronto
Prentice-Hall Hispanoamericana, S.A., Mexico
Prentice-Hall of India Private Limited, New Delhi
Prentice-Hall of Japan, Inc., Tokyo
Simon & Schuster Asia Pte. Ltd., Singapore
Editora Prentice-Hall do Brasil, Ltda., Rio de Janeiro

C·o·n·t·e·n·t·s

List Of Illustrations

"For Karen"

Acknowledgements

The Authors of this book wish to thank the following
for their contributions to the book:

Linda Kay Rader
Rader Gas Consultants
Principal Contributing Author
(Excerpted from *Natural Gas 101*)

Michael L. Kessler
American Energy Solutions, Inc.

Sunil Talati
Parsons Brinkerhoff

Hill and Associates

The McGraw Hill Companies

Section 1:

**Current Happenings in
Electric Utility Deregulation**

Chapter 1
Rapid Changes in the Electric Utility Industry

Deregulation of electric utilities is forcing major changes and creating chaos in a staid industry. It is creating an environment that makes renegotiation of your electric contract a very attractive and realistic choice. Utilities are scrambling to retain or gain market share. New alternatives for power supplies are becoming available, and regulatory agencies are becoming more flexible.

Many end-users are finding the utilities willing to change from a rigid approach to a customer-oriented attitude in anticipation of further changes in the deregulation process. Now is the time to renegotiate your electric contracts!

STATUS OF DEREGULATION

It has been said that a decade from now, consumers will continue to buy electricity and natural gas, but its sellers will look as different from today's utilities as the mall does from the general store.

Deregulation of the electric utility industry, which began in late 1992, is already being felt around the nation. The Federal government mandated deregulation on a state-by-state level. This approach to deregulation differs from that of all other industries, such as airlines, transportation, natural gas and communications. These were basically implemented in all states at the same time. Consequently, progress toward retail wheeling (or "customer choice") in each state has been dependent on relative strengths of those in the state wanting deregulation versus those resisting it.

Progress toward a deregulated environment can be compared to a snowball rolling downhill. It started small, moving slowly at first, but is rapidly moving faster and growing larger. At present, states containing more than one third of the nationís population have set programs to move into deregulation. **Figure 1-1, Full Retail Wheeling Competition**, shows the current status of this effort for each state. The early progress was made in states with high-energy costs, but recently it has spread to states with lower costs, such as Oklahoma. No state wanted to be "first," but the price of being "last" means the loss of business and taxpayers to those states with deregulation.

The model for deregulation specifies that utility companies are to be split into separate generation and transmission-distribution organizations with a "Chinese wall" in between to prevent self-dealing. The generation groups will become unregulated, and will be able to sell their products as a commodity to the highest bidder. The transmission and distribution system will remain regulated to insure reliability and delivery of power to the customers. The generalized model

Full Retail Wheeling Competition

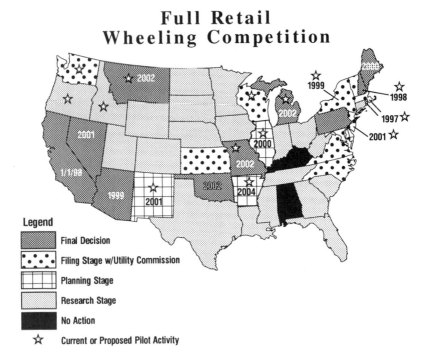

Legend

	Final Decision
	Filing Stage w/Utility Commission
	Planning Stage
	Research Stage
	No Action
☆	Current or Proposed Pilot Activity

Figure 1-1

Future Power Supply System

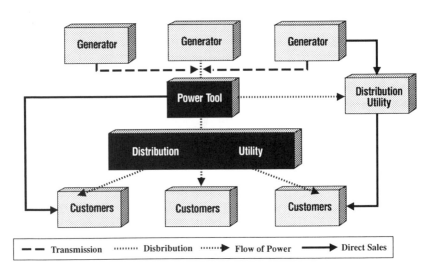

Figure 1-2

of the new utility configuration is shown in **Figure 1-2**, Future Power Supply System.

NEW TECHNOLOGY

Just as in the telecommunications field, new technology is beginning to impact the utility companies' traditional generation patterns. In the 60's and 70's, large central coal or nuclear plants took up to 10 years to build; cost $1,000 – $5,000 per kW; and required 10,000 Btu to generate a kW of energy.

Now, gas-fired turbines take a year to build; cost $300 – $500 per kW; and require 7,000 Btu to generate a kW of energy. These smaller units can be easily located out in the service area, rather than being concentrated in a few central spots. The distributed location of the new units utilizes the transmission system more efficiently.

Better monitoring, controls and communication are allowing the utilities to use the generation more efficiently. Generating units are being more fully utilized.

UTILITY COMPANY REACTIONS

Utility companies' reactions to deregulation vary. Most are still opposing changes. Others are accepting the inevitable, and are taking advantage of the many opportunities to provide new services in new territories. All are taking steps to prepare for a future without the security blanket of the regulatory process. Some of these include mergers, downsizing, new purchasing alliances, management services between companies, diversifying and reinventing.

A respected financial analyst in the electric utility field recently speculated that we are moving toward a nation of super utilities. The current 200-plus companies will be merged to around 20 within the next decade.

One is reminded of the U. S. auto industry several years ago when automakers realized that they were no longer competitive with the Japanese. Some fought hard for trade barriers to protect themselves, while others got busy and markedly improved the quality of their products and their manufacturing proficiency. Now, most of them are quite competitive.

The same pattern is emerging in the deregulation process for utility companies. The adaptable will survive. We consumers must change also, and take advantage of changes resulting from deregulation. It will allow us to break free from the limited rate structure choices that we have endured for many decades.

AGENCIES CONNECTED WITH DEREGULATION

Although the Energy Policy Act of 1992 reserved to the states the right to administer deregulation inside their boundaries (*intra*state), the Federal Energy Regulatory Commission (FERC) is the agency charged with managing the deregulation of *inter*state energy at the national level. It is taking the lead in establishing policy to implement deregulation and to urge movement toward deregulation by the states.

Each state has a regulatory agency, such as a public utility commission, to administer the *intra*state energy activities. These agencies will be the driving force for implementation of deregulation at a state level, with legislative guidance from state legislatures. A problem arises, however, when there is a combination of both *intra* and *inter*state deregulation issues to be resolved.

A further complication is the existence of regional reliability councils, or power pools, (**See Figure 1-3, NERC Regional Councils**) which are part of the North American Electric Reliability Council. They exercise technical control over the

North American Electric Reliability Council Regions

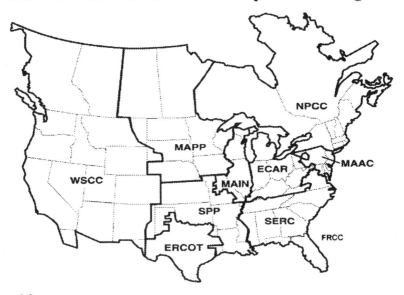

Figure 1-3

utilities within their boundaries, which are generally set at edges of member utility territories. In the future, these reliability councils are the more likely structure to direct much of the deregulation activities within their areas.

Energy users should be aware of the massive restructuring that is well underway within the utility industry, as well as in each of these agencies, as they try to adapt to a deregulated environment. Consumers need to actively position themselves to capitalize on, rather than be penalized by, these changes.

REGULATORY AGENCY ATTITUDES

The state commissioners are currently charged with the responsibility of administering a portion of the deregulation process that is *intra*state. As you might guess, within fifty different agencies, there is a wide variation of attitudes. Many feel that the degree to which the agencies are favorable to the deregulation effort is strongly related to their ties to the utility companies in their states. In some cases, there has been an exchange of personnel that might raise eyebrows. However, most regulatory agencies make a sincere effort to be unbiased in their actions.

Commissions in states with high power costs are the most likely to be aggressive in the deregulation process. The movement in all states favoring some degree of retail wheeling is impressive. The key dates are 1998 and the year 2001. In 1998 a number of additional retail wheeling experiments began. Many think that most of the states will be deregulated by 2001 or 2002. In all states, the legislature has a direct effect on the deregulation process. Many states will begin to see a strong push in the next few sessions of their legislature.

Commissions face an unprecedented challenge in balancing the interests of all parties – residential, commercial and industrial customers, utility stockholders, the financial community and their own staffs ñ in the transition to an unregulated environment. All these parties are jostling for position in this effort.

Many new concepts must be developed and implemented. Court cases will test many of the changes. As a consequence, many state agencies are moving carefully, but are feeling the pressure of the time clock.

Chapter 2
New Choices in the Electric Industry

The deregulation process has some new players that bring fluidity, change and uncertainty. These players will have a monumental impact on the production and sale of electricity.

INDEPENDENT POWER PRODUCERS

Independent power producers (IPP's) have caused major changes in the way electric utilities do business. These independents are more agile, cost effective and less hampered by some of the regulatory constraints of the utility companies. Overheads are reduced. Their new units tend to be smaller and quicker to install, lower in first cost and lower in operating costs. As a consequence, they are very competitive with the utility companies' older, larger plants.

In spite of a considerable effort to hold them down, they are now adding more generation than the old utilities. IPP's offer significant competition that will be helpful for end users in the future by driving down power costs and offering alternative sources of power.

POWER MARKETERS AND BROKERS

A new breed of fast-paced operators is causing more major changes. (See Chapter 8.) They have moved from gas brokering and are now making a market in wholesale power. Power marketers will buy surplus power from one utility and sell it at a profit to a second. The second utility will save money by backing down on high cost generation, or postpone construction of new plants with the purchase of this energy. Power brokers do much the same, but do not take title to the power they handle.

Utilities are selling or retiring their less efficient units and increasing the capacity of their remaining units to meet growing needs. More utilities are buying power from others rather than generating it. As a result, their need for services, such as those provided by brokers or marketers, is increasing.

The future impact of these firms can be compared to the impact that firms like Sprint and MCI had on the long distance telephone industry. Competition resulted in many AT&T customers changing over to a new provider of telecommunications services. In an interesting reversal of roles, power marketers, such as Enron, are now moving to purchase their own utility companies.

AGGREGATORS

One of the techniques that will soon become a strong tool in the negotiating process to reduce power costs is aggregation of loads from multiple sites. These loads will be combined into a single amount large enough to gain some economies of scale, or increase the negotiating clout of those multiple sites.

A power aggregator is a public or private organization that may arrange for electricity to be provided to groups of customers, rather than an individual supply to each customer. This approach makes the load more attractive to a power provider and thus leverages the purchasing power of the customers with the power supplier(s). This accelerates the possibility for customers who use smaller amounts of energy to take advantage of deregulation.

Aggregators may purchase, or arrange for, several blocks of power for multiple customers. The individual load characteristics may be different, but, when blended together, will fit the load characteristics of the aggregated load. This "middleman" role is expected to become more important in the future. Aggregation, as well as many other techniques, will be considered in an effort to provide power at the lowest possible cost to customers (many of whom may be inexperienced at negotiations).

An offsetting concern should be the technical, financial and legal viability and durability of the provider. A number of new companies are jumping at the opportunity to attempt this type of service. However, it will not be easy to make such a program successful for long periods of time.

WHEELING (TRANSPORTING) POWER

A key element to the success of deregulation is the ability to move (wheel) electricity greater distances than has been required in the past. A generating source with lower cost power must be able to sell its output to a variety of purchasers, some of whom might be several utility systems away. Many utility companies have wanted to charge a high fee to wheel power through their transmission systems.

Some relief is now provided by FERC's Order 888, which states that no utility can charge others more for the use of their transmission lines to wheel power than they charge themselves. This prevents the possibility of a major surcharge that would make retail wheeling uneconomical.

There are two types of wheeling charges that are being considered: (1) Postage Stamp Rates that charge the same price regardless of the distance power is moved, and (2) kW-Mile Rates that do consider the distance. Transmission service charges are in the $1.00 – $1.75 range. Based on these charges, it becomes clear why there is a "rule of thumb" stating that it is uneconomical to wheel power more than two utilities away. Alternatively, there is an increase in negotiated or paper agreements which exchange power without actually moving it through several systems.

COMMODITIZATION OF ELECTRICITY AND CONVERGENCE

A very active futures market is another development resulting from deregulation. This addition is patterned after the natural gas futures trading. The new market allows utilities a chance to make more money on their surplus generation at any time by not leaving it idle. They can optimize their generation and power purchasing to lower the overall power costs.

Energy trading and risk management techniques can be used to dampen the effects of the resulting price volatility. They present options that have not previously been available to optimize the pricing of electricity. Price discovery is also quite helpful in planning future management actions. The flexibility of power utilization is greatly improved by (1) hedging strategies, (2) future basis risk, (3) swaps, and (4) other related tools made available by the market at the New York Mercantile Exchange.

Gas is cheaper to move than electricity. Electricity is cheaper to move than coal. There is a strong movement to put them all together as just "energy in different forms," and buy the cheapest form available. For example, plans for a power plant might be to locate it near a coal mine, instead of using unit trains to move coal from a distance to the power plant. The transmission system would move the electricity instead of the coal.

We are moving in the direction of a convergence of energy sources. An end user's requirements for energy can be met from a variety of sources, and the pricing of such sources will approach an equilibrium.

WHOLESALE POWER COST DATA

The United States is divided electrically into regions called reliability councils. The utility companies in each of these councils (or power pools) band together to manage power availability within its borders and the inter-ties with other power pools. Although there were power sales on a wholesale basis among the members of each pool, the pricing of this power was not generally known.

FERC mandated that this information be made publicly available. Consequently, we now have access to wholesale power with pricing tracked on an ongoing basis. This is a useful tool to use in determining what your utility company's marginal costs are for any power purchases. An interesting comparison can be made of this amount and what your power company is charging you.

STRANDED COSTS (STRANDED INVESTMENT)

Easily one of the most controversial and pivotal issues in deregulation of electric utilities is the matter of power supply contracts and/or generating equipment that will no longer be competitive in a deregulated environment. This will occur because cheaper power costs will be available from other sources. In their regulated environment, utility companies sometimes tended to overbuild in the past. A guaranteed return on capital investments provided assurance that these facilities would be profitable.

The regulatory agencies were charged with the responsibility of requiring that all investments be prudent and in the customers' best interests. However, in some cases the judgement of the utilities and their regulators proved to be faulty. Power costs from some generating units – particularly nuclear – are significantly higher than alternative sources. Deregulation is focusing attention on these lapses in good planning. Estimates of investments no longer recoverable in a free market range up into the many billions of dollars, and seem to indicate that the planning and authorization process of all parties did not function as well as possible.

Utility companies take the position that they were encouraged or required by their regulatory agencies to construct these high priced units and, therefore, should be compensated for their resulting costs. They also point to occasions where some of their investments were not allowed in their rate bases by the regulatory agencies.

In contrast to the approach taken during deregulation of other types of businesses, utility companies are taking the position that they should be compensated for all uneconomic investment. In addition, they want to be compensated for the costs to convert to a deregulated condition. Industries such as airlines, telecommunications and natural gas generally wrote off any uneconomic investments during deregulation.

Merely writing off these investments is vehemently opposed by many of the utilities. Rather than impacting their stockholders, their preferred alternative is to pass these costs on to their customers. Those who exit their system in order to buy cheaper power elsewhere are to be charged a lump sum for their share of this investment.

Utility companies are making an extremely strong effort to be compensated for all of their investments that will no longer provide them revenue. Currently being debated is one mechanism to recover stranded costs for the utility companies called "securitization." It will allow a utility to recover its stranded costs *up front* in one lump sum payment from the sale of specialized bonds. The definition of this term is "converting into marketable securities" (securitizing) the present value of the revenue stream anticipated to be produced by customer payments of stranded cost recovery surcharges over a period of time (five to ten years). Under this plan, the legislature (or utility commission) irrevocably orders that customers must pay a surcharge as part of their electric bill to complete the bailout of the utility having stranded costs.

Since the bonds are likely to be favorably rated, they will bear interest at rates less than the utility company's other borrowings, and some of the proceeds of the stranded cost recovery bonds can be used to pay off pre-existing debt. This will result in a token reduction in electric rates. However, electric rates would be reduced an even greater amount, and years earlier, if customers could buy electricity in an open competitive market without paying stranded cost recovery surcharges in the first place.

Many say that there are strong indications that some utilities are inflating their estimate of stranded costs, now estimated at $135 billion. By comparison, the gas industry originally estimated their gas restructuring costs at $44 billion. The final number was approximately $13 billion. These costs were finally allocated between pipeline companies and gas consumers.

There appears to be little interest among utility companies to follow this pattern set by natural gas companies. Many utilities prefer to delay the advent of retail wheeling until the stranded costs can be depreciated in the usual manner.

This battle is ongoing, and may have a significant effect on customers' ability to choose a supplier other than the local utility. Customers should take an active part in the deliberations to establish the ground rules on stranded costs, or Competitive Transition Charge (CTC), as it is sometimes euphemistically called. Others have rather harshly called it the "price of stupidity."

Customers should try to structure negotiating efforts to minimize the impact of stranded costs on future actions. Negotiate to allow as much flexibility as possible.

UTILITIES WILLING TO NEGOTIATE NEW CONTRACTS

For the first time, utility customers are moving from a captive status to having a choice in suppliers. Utility companies are beginning to understand that the day is coming when (like private industries) they must rely on price, good service, customer relations and other benefits to keep their customers. The specter of competition is forcing them to make concessions that were unheard of just a few years ago. Many utilities are attempting to realign their relationships with customers quickly before they lose them to other suppliers.

SECTION 2

Preparing for Successful
Electric Negotiations

Chapter 3

Start Your Savings Now

The intense contest for profitability now faced by many manufacturers and commercial firms is no longer local. It is national and even international. Competition in business has been a continuous, overriding concern for many years; competition for utilities is just beginning. Winds of change are blowing strongly for electric and gas utilities. The massive deregulation programs that are underway offer exceptional opportunities to restructure your relationships with utility suppliers.

WHY NEGOTIATE NOW?

Now is the time to start negotiating with your utility company for an improved package of rates and other benefits. The pressures of the coming deregulation make most utilities much more willing to talk. If some benefits are developed that can be offered to them in return for a more favorable rate package, you significantly increase your negotiating power. The following examples illustrate several results of successful recent utility contract negotiations **(See Figure 3-1: Negotiations Equal Results.)**

Negotiations Equal Results	
Negotiations	**Results**
Midwest Manufacturer	25% electric rate reduction
Midwest Cement Plant	22% electric rate reduction
Northeast Steel Company	25-30% reduction translates into "millions per year"
Northeast Metals Firm	20% savings – provides expansion possibilities
Northwest Manufacturer (computer parts)	$250,000 per year saved with "customer choice"

Figure 3-1

Those not familiar with the utility industry tend to assume that they must wait until full deregulation occurs before they can renegotiate their utility contracts. Actually, many basic concepts of "Customer Choice" in deregulation, as it is now called, are already accepted. The focus is on a scramble by all players for advantageous positions for the future. A number of states are beginning, or have scheduled, pilot programs to start the deregulation process.

Most utility companies are further along in this restructuring process than many of their customers. They realize that they must reshape their image from condescending monopolies to friendly partners and realign their relationships to assure that their customers will remain tied to them for many years. They are also reaching out to new markets in their existing territory and eyeing neighboring utilities' customers.

Most of these utilities are reluctantly accepting the need to shift from tariffed, one-size-fits-all rates (approved and protected by the state regulatory agency). They are switching to bilateral contracts that more accurately reflect their customers' needs. Be aware that much of this shift is being grudgingly revealed to their customers.

Major energy users now have a wide range of both incentives and valid threats to use in moving utilities to the negotiating table. The trick is to use these tools successfully in the negotiating process.

It may take time to wade through all the technical, political and financial issues involved to take full advantage of the opportunities for savings that currently exist. It is well worth the effort, however, when one considers that the benefit to cutting the cost of power, in addition to cutting the consumption of power can lead to immediate savings. Paybacks from these savings can be measured in weeks or months instead of years.

DO YOUR HOMEWORK
Advance Work, Carefully Done, Makes the Difference.

Probably the most important piece of advice in this book is contained in the previous sentence. Keep in mind that the utility companies specialize in negotiating utility contracts of all types, with all types of consumers. They are expert at this process. However, most of their customers have little experience in this field. As a result, they are prone to oversimplify the process and get too impatient for quick results.

You need to develop a thorough understanding of:

- Your own power and energy use patterns
- Possible future change in these patterns
- Possibilities of internal flexibility in the way your power is used
- State legislative climate
- State regulatory climate

- Regulatory Council or Power Pool regulations
- Contract terms given to other customers
- What the utility company really wants
- What you really want from your utility contract
- What you are willing to give in return

Take the time to do your homework and lay a proper foundation for the negotiation process. Trying to shift directions in mid-negotiations may be quite difficult.

Gather Information on Your Utility Company

All utility company web pages on the Internet provide a wealth of information. In addition, selected portions of a company's 10K's can be downloaded from the SEC web page. It provides more detailed and less market-oriented information than annual reports.

Much more helpful, but potentially quite expensive, is information on generation and transmission costs, rates, comparative financial conditions and potential stranded investment costs. This specific information is available from a few selected utility databases. The cost of this information is in the several thousand-dollar ranges, but is well worth the expense.

Build a Computer Model of Your System to Test Proposals

One of the most important steps in the negotiation process is the preparation of a computer model of a facility's operation. This will help test various combinations of rates and procedures such as load-management, load-shifting, real-time pricing or similar alternatives. For maximum effectiveness, the model should be constructed on an hourly basis. Much of the data can be obtained from the utility company. Other input must be estimated from daily or monthly data.

Benchmark the model against the most recent year of operation (at least), and adjust until there is an acceptable match. Sometimes the negotiations will move rather quickly, and you must be able to determine the impact of the latest offer on your operation in a short period of time.

Have a Knowledgeable Utility Attorney in Your Pocket

Chances are that your new contract will become much more complicated than those of several years ago. It should also be much less one-sided than the older one. The new contract may cause the utilities to raise questions about its acceptability by the regulatory agencies. Often, a utility negotiator will say that a clause you want included is not legally possible or acceptable to the regulators. Your attorney can be a big help in a limited role, although they sometimes do not understand your utility requirements and might not be as effective in a direct negotiating posture.

Coordinate Carefully with In-Plant Changes

As you determine the direction of the negotiations, you can evaluate modifications to your overall utility management strategy to take advantage of the new situations. This can be a powerful means of reducing utility costs in your facility.

Chapter 4
Develop Negotiating Alternatives

It is unrealistic to expect utility companies to give up their privileged rate position without a serious fight and strong justification on your part. Your negotiating team should develop some realistic alternatives in an attempt to balance the negotiating field.

Strongly consider some weapons *(Clubs)* and some positive incentives *(Candy)*. Build a pile of *candy* and a pile of *clubs* to use in your negotiations. Following are some of the choices that might be explored.

CLUBS
Reduce Your Load

Plant Shut-Down
Threaten a shutdown of your plant. The only thing that would be worse to a utility company than reducing your rates would be to lose your total load. If you can make this argument really stick, it is a very powerful one. **However, bluffing is tough to do.**

Shift Current Production Site
Threaten a shift of part or all of your production to another site. This might reflect a severe reduction of manufacturing plant operations with a consequent reduction in power load.

Shift New Production Site
Threaten to put a proposed new production at a remote site. This is a very powerful argument, because you are offering a new load to the utility in return for rate relief.

Look at Different Power Sources

Convert to a Municipal System
Work with your municipal government to consider municipalization of your local utility system. Some cities are taking over the electrical distribution system of the investor-owned company supplying power within their city limits. Then they can purchase cheaper power from alternative sources on a wholesale basis. The legal and political hurdles of this alternative may be severe.

REA Cooperative
Purchase your power from a REA Cooperative. Some co-op's are now willing to work through the legal, regulatory and political hurdles to start supplying power to utility company customers if your plant is in a dually certified territory.

Power Marketer

Purchase your power through a power marketer. One of the best ways to make the above two alternatives really valuable is to have a marketer offer cheaper power to your new source. At present, this would have to be done on a wholesale basis to the municipality or co-op. Consider using a RFP for power, and let the respondents develop creative ways to serve you.

Buy Down Your Contract

Sell the contract you have with your utility company to a third party. Some are willing to make regular payments to you before deregulation in return for the right to supply power to you when deregulation is complete. These payments could last for several years until you have *customer choice.* In effect, this lowers your cost of power.

Local Generation

On-Site Generation

Install an on-site generator (a diesel unit or a more efficient gas turbine-driven generator) that will supply all, or a significant part, of the power requirements for your facility. As an alternative, it might be used to shave the peaks off the demand, if you have high demand charges. This option may be tough to justify unless you have a low-cost fuel source, or can use the waste heat in your processes. Many companies do not want to have the responsibility of operating a power plant. It might be possible to turn it over to a third party to run.

Emergency On-Site Generation for Utility

Dedicate the on-site generation to the utility company's use if they have an emergency. If you have a backup generator, or choose to add one, some utilities will contract with you to consider it an emergency source of power for *them,* if they can dispatch it. They may pay you handsomely for this privilege.

Merchant Plant

There are many organizations that are willing to install a larger plant on your property and sell you part of its output and market the rest. This usually gives lower costs than the previous two alternatives.

Does It Make Sense to Buy Your Substation?

Are you metered on the high-side or low-side of the sub-station? If you own the substation, you probably can get a lower transmission voltage rate. The offsetting cost of ownership and maintenance must be considered. Typically, most utility companies will now allow you to purchase the substation on reasonable terms. Remember that utilities usually depreciate their equipment on a 30-year basis, so go for the depreciated value.

Most utility companies will provide maintenance services for a fee. Because sub-station maintenance can be a real hassle it is well worth the cost to have experts handle this function.

An important long-range factor is retail wheeling. Ownership of your own substation could give you added flexibility to take advantage of wheeling from others in the future.

Look Inside Your Plant

Internal Peak Shaving

Reducing peak demands is one of the most beneficial means of reducing power costs. After an initial reaction of "there are no opportunities to cut any operating loads during peak periods," the staff can usually make some very significant cost reductions with this approach. Keep attention levels high, and let your staff be creative with this one.

Internal Energy Use Reduction

Most people think about reducing internal energy use when they are trying to cut power costs. Usually, there are some significant opportunities in shutting off unnecessary equipment or reducing the load on some of the equipment. Again, the workers will usually have the best ideas in this category.

Utility Technical Help

Ask your utility to provide you technical help from *EPRI* or other sources. Look at the big picture. Ask for everything that you think might be a possibility. Most utilities can provide you some technical assistance from one of their trade organizations or from internal sources. **Don't be bashful!**

CANDY

Create Win-Win Partnerships

The idea of *Win-Win* is a rather unusual one in utility-commercial relations. Usually, the best reaction one could expect to the concept was "lip service" from a utility company with captive customers. As a result, the stage was set for an antagonistic, *"us versus them"* type of relationship.

Many customers and utility companies agree that they want to move beyond that narrow approach to a true partnership arrangement. However, accomplishing it is not an easy task. There are strong challenges in changing the way utility and customer executives view each other, but it is worth the effort.

Contract Length

The impending massive changes brought by retail wheeling sets up strong conflicts on the term of the contracts. Users want short terms, while utilities want long terms. Given normal conditions, you should opt for no more than a three-year length in your contract. However, if you can extract sufficient concessions and provide flexibility to take advantage of retail wheeling when it comes, a longer length contract can be a significant attraction to your utility company.

Load Management to Match Their Needs

Utility companies, with few reserves of power, want the highest possible load factor in their customers. This provides the most revenue for them, with the least amount of generation equipment. Serious efforts to smooth out peak demands may be very attractive to them.

Purchase Other Goods or Services from Them

Utilities are trying desperately to diversify, and are looking for other sources of revenue beyond their high-priced generating facilities. Many are offering various specialties such as security, heavy equipment testing, or engineering services. They will sometimes provide performance contracting on in-plant equipment upgrades. These additional services may be beneficial if you have a need for their specialized expertise. Otherwise, be very cautious about lack of experience and high overhead factors.

One of our clients buys ash that is produced by their utility company power plant. They use the ash for raw materials in their own manufacturing process, and also provide waste disposal sites for some of the ash.

Offer Other Facilities' Loads

If you have other facilities that are located a reasonable distance away, it might be possible to aggregate your loads. The greater the load you represent, the more attention you can get from your utility company. Aggregating, or pooling loads, may not yet be possible in your area. If not, just negotiate together for separate contracts.

Chapter 5
Risk Management

The old approach of a standard rate for everyone provided manufacturers a cocoon of protection from choice and responsibility for utility rate management. It was also quite restrictive and expensive. With rate relief there are many more choices to make and an increased possibility of making wrong decisions. Typically, the greater rate relief for a customer will come from assuming greater responsibility for managing your utilities.

A significant question to be answered concerns the amount of risk of utility price variation, and management's acceptance of this risk in the future. Considerable education may be required to get them to see that normal business risks now extend to managing utilities; however, the potential benefits usually greatly outweigh the potential hazards.

One major new development is risk management procedures, similar to those procedures used in the natural gas arena. Hedging, puts, calls, caps, floors, collars, spreads, etc. in the power supply arena are going to be commonplace in the near future.

Associated with any kind of freedom are some corresponding risks. This is inherent in any new ability to make choices. Some of them may be costly.

This truism is certainly visible in the deregulation of wholesale and retail electricity and gas markets. In the past, there was no alternative but to accept a product at a uniform price set by a joint collaboration of the supplier and its regulatory body, without regard to its value as set by the market. There is a growing opportunity to let that market set the price.

This is an effective but messy way of determining the true worth of a product in the minds of its users. Add to this scenario the declining margin of spare generating capacity caused by a combination of continuing load growth across the nation. Couple this with a strong reluctance by utility companies to build new generation due to the uncertainties of the future. Similarly, the deregulated environment probably will force suppliers to attempt to operate with slimmer generating margins. Another factor is the aging of many nuclear power plants that are nearing their retirement age.

The combination of these factors will tend to cause a significant volatility in the electric market. On a recent day in the Midwest, electricity spiked at $7,000.00 per megawatt hour ($7.00 per kWh), up from its usual $15 to $25 per megawatt hour ($.015-$.025 per kWh) trading range. This was a frightening event, but its duration of several hours was brief when compared with the prices of the remaining 8,700+ hours in the year.

NATURAL GAS PRECEDENTS

As natural gas made the transition to market-based pricing through deregulation, that market became more efficient internally, but price volatility increased. As a result, natural gas prices have been among the most volatile of global commodities. Prices of electricity are expected to be much more volatile.

Natural gas prices are not as sensitive as electricity prices. Neither does delivery pose as large a problem for natural gas, while it can be a major problem for the delivery of electricity. Natural gas can be stored. Large gas storage facilities across the country work to reduce many delivery problems. Electricity, however, cannot be stored in a cost-effective manner. This is another factor that further increases the volatility of the electricity market.

Since the nation is much farther down the path of deregulation of natural gas, a robust market for natural gas has been developed to manage risk and supply. This market is serving as a model to be used in developing one for the electrical market.

Numerous natural gas contracts are actively traded on the New York Mercantile Exchange, and the prices for this product in various parts of the country are relatively well established for several years in the future. This is in spite of significant near term price variation. By contrast, electricity futures contracts go out 18 months, and only the first three or four monthly contracts are actively traded. Consequently, the future prices for electricity are not as reliable as future prices for gas.

One of the greatest challenges facing executives is how to manage the risks associated with market-based pricing of electricity. Managing risks will be a constant necessity in the future. Some risk management tools used extensively for decades to manage risk in commodities are being adapted to the electrical arena. Risk management and futures contracts are quickly becoming a fundamental part of the electric power business. Just as quickly, dealers are becoming aware that the techniques that are used in financial markets are inadequate for the energy arena, because of its complexity.

As Dragana Pilipovic states in her important book, *Energy Risk**, "What makes energies so different is the excessive number of fundamental price drivers that cause extremely complex price behavior. This complexity frustrates our ability to create simple quantitative models that capture the essence of the market. A hurricane in the Gulf of Mexico will send traders in Toronto into a tailspin. An anticipated technological advance in extracting natural gas could be influencing the forward price curve. How would you go about capturing these kinds of resulting price behaviors into a quantitative model that is also simple enough for quick and efficient everyday use on the trading desk?

In the same book, specific differences are shown in Table 5-1*.

*The above mentioned material is reproduced with permission of
The McGraw-Hill Companies. *Energy Risk* © 1997, by Dragana Pilipovic.

What Makes Energies Different?

Issue	In Money Markets	In Energy Markets
Maturity of market	Several decades	Relatively new
Fundamental price drivers	Few, simple	Many, complex
Impact of economic cycles	High	Low
Frequency of events	Low	High
Impact of storage and delivery; the convenience yield	None	Significant
Correlation between short- and long-term pricing	High	Low "split personality"
Seasonality	None	Key to natural gas and electricity
Regulation	Little	Varies from little to very high
Market activity ("liquidity")	High	Low
Market centralization	Centralized	Decentralized
Complexity of derivative contracts	Majority of contracts are relatively simple	Majority of contracts are relatively complex

Table 5-1

WHY ARE ENERGIES DIFFERENT?

Another important requirement of a competitive power market is market liquidity. Accessible means of buying and selling power is vital, and price ranges must be easily determined. Futures can play many valuable roles in this effort, such as addressing price volatility, facilitating price discovery, enhancing marketing flexibility, and providing systems safety and anonymity.

Although it is likely that few end users will be active in trading electricity, they are becoming more cost and value-conscious. Therefore, they are more receptive to a structured approach that provides them an opportunity to both reduce costs and hedge against unforeseen price increases.

RISK MANAGEMENT DEFINITIONS

Various financial tools give participants in the marketplace the capacity to affect the outcome of risk exposures to undesired volatility of future electricity or gas prices. Fundamentally, steps are taken today to reduce the chance or risk of a negative pricing event in the future. This can be somewhat compared to buying an insurance policy against some undesirable future event.

The term "risk management" is often used interchangeably with the term "hedging," although "risk management" covers other components, as well. Three key elements of hedging are 1) futures contracts; 2) derivatives; and 3) basis.

Understanding the term "basis" is very important when hedging. Basis is the relationship between the price of a cash position on a product being hedged and

27

its futures contract price. If a futures contract is being used to hedge an exposure with terms different from one that is identical to the futures contract, the basis will fluctuate. "Basis risk" is an undesired change in the basis. Delivery causes the basis to disappear as cash and futures prices converge at the expiration of the futures contract.

A derivative instrument is one whose value depends on (derives from) something else. The "something else" can be anything of value such as a commodity (agricultural product, natural gas or electricity), index, or security. Derivative trading involves the exchange of rights or obligations based on an underlying product, but derivatives themselves do not directly transfer property. They are used to manage the financial risks associated with the underlying product.

Derivative contracts are divided into two categories: Forwards and Options. Forward-based products include Forwards, Futures and Swaps. Option-based products include 1) Options, 2) Caps, 3) Floors and Collars, 4) Hybrids, 5) Options on Futures, and 6) Forwards and Swaps.

- A *forward contract* commits the buyer to purchase, and the seller to deliver an asset at a specified time in the future at a prearranged price.

- *Futures* are standardized forward contracts that are traded on organized exchanges.

- A *strip* is a series of consecutively maturing futures contracts, allowing the purchase of a futures contract for each month in the period in one transaction.

- A *swap* is a series of consecutively maturing forward contracts. Two parties agree to exchange a series of cash flows based on a liability or asset. *Basis swaps* allow participants to exchange two floating payments based on different indexes.

- *Spreads* are similar to swaps except that the payment streams are tied to different products. *Spark spreads* would involve the exchange of payments tied to the price of fuel used in generation for payments tied to the price of electricity.

- *Options* are contracts that give the buyer the right, but not the obligation, to purchase or sell the underlying asset at an agreed-upon price in the future. *Options buyers (Longs)* pay *sellers (Shorts)* a premium for this right.

- *Call options* give buyers the right to buy the underlying asset from the seller at the prearranged strike or exercise price.

- *Put options* give buyers the right to sell the underlying asset at the exercise price.

- *Options in-the-money* are when the strike price is less (in the case of call options), or more (in the case of put options), than the market price of the underlying asset, and ones that can be exercised at a profit. Options out-of-the-money would be allowed to expire unexercised.

- *Caps and floors* are a series of consecutively maturing options priced as a single contract. A *cap* is a series of call options and a *floor* is a series of put options.

- A *swaption* is an option on a swap. The purchaser of a *swaption* is buying the right, but not the obligation, to enter into a swap at the exercise date on prearranged terms.
- A *collar* is a combination of a long cap and a short floor. Risk managers sometimes fund the cost of acquiring a cap by simultaneously writing a floor.

(Note: See other definitions in the Glossary)

Only a small part of traded futures contracts result in the actual delivery of product. One power transfer may be the subject of multiple trades. Some power marketers are accused of excessive trading to increase their apparent volume of power sales.

The last decade has seen dramatic growth in new varieties of off-exchange instruments. Swaps and related options are rapidly growing segments of the "derivatives" family. They are not traded on a futures exchange, but are privately negotiated to meet the particular risk management needs of the parties involved.

HOW RISK MANAGEMENT IS USED

Risk perception is based on some common factors that include the probability of the negative event, catastrophic consequences, previous event experiences and controllability of the event. The most effective approach for reducing perceived risk is to give the individual more control over its consequences. Thus, risk management techniques are becoming more heavily used.

Derivatives are a low-cost way of transferring risk from those who would rather not be exposed to it, to others who are either able to handle that risk or who bear opposing risks. Popularity of options occurred when it became apparent that they provided insurance protection from adverse price movement at a reasonable fixed cost. The cost of buying options to hedge against adverse price movement is a cost of doing business just like any kind of insurance.

Risk management is used to:

- Protect prices and costs
- Maintain margins
- Provide a competitive advantage
- Afford more confident budgeting and planning
- Allow "monetization" of poorly performing assets
- Temper earnings volatility
- Focus corporate strategy
- Enhance a firm's ability to borrow

In the past, regulated utilities have been motivated to develop sophisticated purchasing strategies, either for their fuels or for purchased power. With the advent of deregulation, purchasers are becoming more aware of relative values of these products. The advent of the power marketers has brought a movement toward a broader use of these pricing strategies.

RISK MANAGEMENT MECHANICS

The main forces driving fluctuation in power prices include:

- Generation reserve margins
- Fuels price and availability
- Power plant operating costs
- Power line disturbances and wheeling concerns
- Temperature and freak weather conditions
- Daily and seasonal demand curves
- Plant outages or strikes
- Energy regulation
- Market liquidity
- Control system problems
- Environmental regulations

Futures contracts are governed by the Commodity Futures Trading Commission (CFTC). Futures exchanges are backed by well-capitalized financial institutions. Thus, firms who use futures can feel comfortable with the financial stability of the transactions.

Some derivatives are traded only on an organized futures or options exchange, while others [off-exchange or over-the-counter (OTC) derivatives] are not. Exchange-traded instruments include all futures contracts and all options on futures. All forward contracts, swaps, swaptions, hybrids, caps, floors and collars, as well as some options are traded outside of a formal exchange.

FORWARD PRICE CURVES

There must be some degree of comfort in the future prices of electricity for the risk management process to be healthy. Since electricity price volatility is affected by so many factors, predicting the future numbers is quite precarious. A forward curve of realistic estimates of future power prices in a non-cash-and-carry market such as electricity is very much needed to identify arbitrage opportunities, to preserve internal consistency and, most of all, for risk management.

After a somewhat slow start, trading is likely to grow very quickly in the coming years. This will occur as deregulation becomes a reality in more areas and trading capability becomes essential to profitability and risk management. As a result, the development of forward price curves will serve as the foundation for all product pricing and risk management.

Most available energy prices today reflect only the current regulated supply and demand rates, not future rates. Predicting the future power prices from the past is quite risky.

Price prediction can be loosely divided into two categories: near-term fluctuation over a period of several days up to several months, and long-term fluctuation over periods of up to ten years (possibly). Some experts believe that realistically accurate price predictions cannot extend beyond a few days.

Further division includes the use of various models to predict the future, or prices fixed by a contract for delivery at some time in the future. Obviously, the latter alternative is the most accurate. However, the number of such contracts is quite limited and many are confidential. A growing contingent of power marketers and utilities base contracts on indexes from published price surveys, such as Megawatt Daily's. See Chapter 6 for more details on longer term pricing projections.

RISK MANAGEMENT STANDARDS

Derivatives for risk management can result in great exposure (fluctuating more widely than the products they represent), in markets characterized by volatile exchange rates, interest rates and commodity prices. However, not hedging with derivatives may be a more risky strategy than using them.

A Risk Management Program needs to be directed by high-level corporate officers, because there will be many opportunities for financial overexposure. Use knowledgeable senior staff for guidance.

Risk management activities, with the accompanying internal controls, are expensive to put in place and maintain. Utilities, power marketers and brokers will certainly have sufficient traffic to justify such a program. End users may be better off to use the services of some outside agency.

RISK MANAGEMENT IN THE FUTURE

Within five years the electricity derivatives market will likely develop into one of the largest of commodity markets. Participants will be the utility industry, power marketers, end users and speculators.

The utility industry is divided into two camps: those willing to implement hedging strategies to soak up their risks and those running for regulatory cover. They are entering an environment in which price risk is a key factor that may be difficult to pass through to the customer.

The two primary characteristics that will determine the winners and losers in the new environment of energy supply are (1) financial risk management skills and (2) marketing skills of new integrated utility suppliers, marketers and traders. Expect major positioning efforts in the future - some of which will be successful and others will not.

For the end user, purchased power contracts that provide risk management services and power are growing. As wholesale competition increases and transmission access is opened up, the demand for derivative-based contracts will continue to expand. End users should push for as high a degree of risk protection as possible.

This protection can be delegated to your power suppliers, who should have a base of expertise built from their activities in the power field. However, the days are thankfully approaching an end when one is forced into major decisions by default on power sourcing and pricing to suppliers.

End users should plan now to develop some degree of capability separate from their suppliers. Start small and build as experience is gained, or seek assistance from an outside third party expert.

Chapter 6
Forecasting Future Electricity Prices

SHIFTING FROM FIXED TO VARIABLE PRICING

In the past, end users had little concern about future power prices. The regulatory process allowed only limited input from them in any intervention effort during infrequent rate hearings. Now, however, the nation is moving toward the time when the rates for power supply (if not transmission and distribution) will be set by the market. Utility companies, power marketers, financial institutions and end users are suddenly faced with predicting the future in power prices with little from the past to guide them.

These projections of future prices, known as the forward market or forward curves, are vital for asset valuation, risk management analyses, new generation planning, deal analysis and contract negotiations.

MARGINAL COSTS (PRICES)

Currently, most power prices in the nation are based on regulated "cost-of-service" determinations that use average costs (total costs divided by total sales). In a deregulated environment, prices will be established by the competitive forces of the marketplace based on the marginal cost of producing the power. In a perfect free market, these prices will tend to move downward to the marginal cost of producing the next required kilowatt-hour of electricity during each hour of the day. In a normal, free market, prices will vary either side of this marginal cost, depending on the strength of the demand for the electricity.

Marginal costs are the operating and maintenance (O&M) costs of the last and most expensive power plant required to meet the system load. In periods of high demand, prices may rise above the marginal costs of that latest unit to provide some return on capital investment. In periods of low demand, prices may even fall below the marginal costs to produce that power.

PROJECTION TECHNIQUES

There are three general types of techniques to develop projection of future conditions:

1) Historical projections – Extension of past data to predict future trends

2) Econometric models – Use of data on a macro scale, using data such as GNP projections that yield generalized results

3) Linear programming models – Use a large quantity of data specific to a given field that produces focused results

A number of projections of future power costs have been made by various organizations. Some of the better ones were done by Hill & Associates, Resource Data International, IREMM, and the Energy Information Administration within the Department of Energy. Each one approached the task of predicting future power costs for different sections of the country established by the National Energy Reliability Council (NERC), as shown in **Figure 3-1 (NERC Regional Councils)**. They used large amounts of historical operating data on:

- Load projections
- Reserve margins
- Power plant efficiencies and marginal costs of production
- Projected fuel costs
- Projected retirements of older units, including nuclear
- Estimates of additional generation that would be required to meet the load
- Transmission system constraints
- Transmission line losses
- Maintenance schedules
- Stranded cost estimates
- Environmental regulations

The Hill & Associates report projects the power costs for each of the four seasons of the year – spring, summer, fall and winter, as well as on-peak, off-peak and spike periods (4:00-7:00 P.M.). An example of marginal power costs for ERCOT is shown in **Figure 6-1: Marginal General Cost Forecast.**

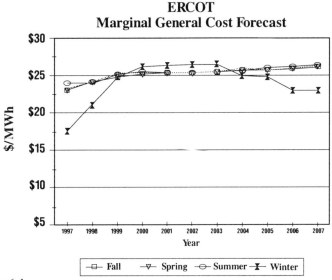

Figure 6-1

The impacts of the variables analyzed were quite different for many sections of the nation. These differences reflected the various mixes of generation types, fuel sources and transmission constraints.

The results also varied between the study authors. A comparison of the different analyses for the AEP territory is shown in **Figure 6-2: Comparison of Power – Projected AEP Marginal Power Cost.**

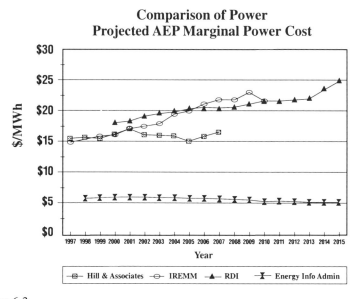

Figure 6-2

Those prepared by EIA are available on the Internet. Those from Hill & Associates, RDI, or IREMM, must be purchased from them. They are valuable in planning strategies for negotiating with utility companies, and for checking any proposal for power pricing that is offered.

FUEL COST PROJECTIONS

A major determining factor in the analyses was the projection of fuel costs. Coal supply prices are predicted to decline in the future. Similarly, coal transportation costs should also decrease slightly. Gas prices were variously projected to be slightly higher or lower in the future, depending on the study author. Summer gas prices probably will be lower than those during the rest of the year. Other fuel costs, such as nuclear, likely will remain fairly constant.

ANALYSES CONCLUSIONS

Some of the general conclusions these studies set forth include:

- Competition, productivity improvements, improved heat rates and efficiency improvements should reduce power costs.

- With full-scale competition in generation services, retail prices for electricity could be lower by as much as 6 to 13 percent within two years. The price changes would vary from region to region, and some regions could even see greater price increases. Under intense competition, prices could fall by as much as 24 percent.

- Some utility companies could experience substantial reduction in market value and shareholder stake due to these price reductions unless they aggressively cut costs.

- Stranded or uneconomic costs related to units or contracts no longer economically viable in the deregulated economy could range from $70 billion to $150 billion. This amount will decline in the next few years as customers are forced to pay down these uneconomic investments.

- Marginal costs in the summer are projected to increase gradually over the next ten years in all NERC regions. In the spring, fall and (sometimes) winter seasons, prices are expected to decline, or at least hold steady.

- Marginal costs will be tied more directly to the price of natural gas.

- Utility profitability in many regions will decline, but most utilities will have positive variable operating margins.

- Non-utility generators (NUG's) cause a disproportional share of the stranded costs. They supply 7 percent of all electricity to the grid, but they represent nearly 30 percent of the stranded costs.

- The single weakest link to an effective, fully deregulated market will be the nation's electrical transmission system. It is not designed to move power great distances beyond the territory of each electric company.

Chapter 7
Choosing Pricing and Terms

PUSH YOUR UTILITY COMPANY FOR NEW RATES

Carefully evaluate your facility's needs, then consider the options to meet them. You must be willing to weigh all the risks against the benefits to find the rate, or combination of rates, that best fits your situation. Many utilities still have only a limited variety of rates. A gentle-to-firm push may be appropriate to get a rate package more suitable for your needs.

Some of the basic alternative rates worth considering include:

• **Time of Day**

Use a Time of Day rate for special applications that allow reductions in normal load during peak daytime hours. Carefully examine your manufacturing process to evaluate shifting part of your load to off-peak periods. This requires strong coordination with manufacturing, but will cause little objections from the utility companies who will gain from flattened patterns of electrical use. The resulting benefits to customers can be quite attractive.

• **Real-Time Pricing**

The real-time pricing rate is a major step in the direction of spot market prices in a deregulated environment. This alternative varies the price you pay for power each hour of the day, based on the utility's marginal costs. As shown in **Figure 7-1, Price-Duration Curve**, there will be some hours during the year when prices are quite high (several times the average rate you are paying now).

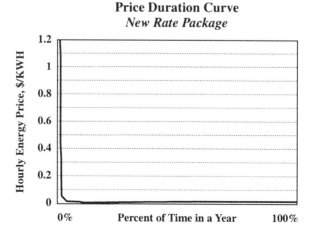

Figure 7-1

During the vast majority of the time in a year, however, the price will be much lower. You must be aggressive in managing your peak loads to get the most benefit from this rate. There are a number of variations in the structure of this rate, a few of which are more beneficial to customers. The ability to manage loads – reduce them during peak demand periods – can greatly increase the benefits from this rate.

- **Interruptible Rate**
  Use an Interruptible rate to lower demand costs. Many plants can allow a part of their load to be curtailed in return for attractive reductions in the demand charges. Production staffs are often reluctant to agree, so careful study, followed by some internal *give and take* will help reach a balance that will optimize the profits to the plants.

- **Interruptible Rate with Buy-Through**
  If you like the advantage of an interruptible rate, but the interruptions cause major problems in plant operation, consider a buy-through clause in your contract. The buy-through will eliminate any interruptions to plant operation. This technique would allow your utility, during a curtailment, to purchase your required power from the grid, then transfer it to you. This temporary power source will be more expensive than your normal power, but its short duration should still make it quite attractive.

- **Interruptible Rate with Buy-Through and Limited Interruptions**
  Another alternative is to use an interruptible rate with a buy-through and limited interruptions. Combining the two previous recommendations might work for some manufacturers – a favorite of a Wisconsin utility.

CONSIDER INDEX PRICING

It is impossible to predict exactly what changes will occur in the future pricing of power and energy. Similarly, during periods of high profits it may be easier to pay higher prices for your power than during those cyclical periods when product pricing drops. Sometimes it is appropriate to tie future rates to some index, (such as the price of your product) in order to better balance plant profitability. To increase the liquidity of power marketing, the New York Mercantile Exchange has established a very active market in power futures contracts – an activity destined to have a growing impact on the price and availability of power across the country. A tabulation of recent prices for on-peak and off-peak power across the nation is shown in **Figure 7-2; Recent Prices of Spot Electricity.**

Look at both of these indexing possibilities for providing future flexibility. Keep in mind that this market is quite young and may evolve significantly in the coming months.

Prices Of Spot Electricity/MWH
(Week Ending February 21, 1998)

Location	Weekly Range On-Peak	Weekly Range Off-Peak
Western Markets		
COB/NOB	$16.25 – $21.50	$9.50 – $13.25
Four Corners	$16.50 – $23.00	$10.00 – $12.75
Palo Verde	$12.50 – $15.25	$10.50 – $12.50
Mid-Columbia	$14.50 – $18.00	$9.50 – $13.50
Midway	$23.50 – $29.00	$16.50 – $19.50
Northeastern Markets		
PJM	$17.25 – $20.50	$12.00 – $15.00
East New York	$23.50 – $29.00	$16.50 – $19.50
New England	$24.00 – $27.75	$17.00 – $20.00
West New York	$18.50 – $21.00	$13.50 – $16.25
Midwestern Markets		
Ameren	$16.25 – $18.00	- N.A.-
Cinergy	$14.75 – $17.00	- N.A.-
CobEd	$17.75 – $19.75	- N.A.-
ECAR	$14.75 – $17.00	$11.50 – $13.50
MAPP	$15.00 – $18.00	$9.00 – $11.00
MAIN	$17.00 – $19.75	$10.75 – $12.75
Southern Markets		
Entergy	$17.00 – $18.75	- N.A.-
ERCOT	$20.00 – $23.00	$11.00 – $12.50
Florida/Georgia Border	$17.50 – $18.75	$13.50 – $15.00
North SPP	$17.00 – $19.00	$11.00 – $14.25
SERC (w/o Florida)	$16.25 – $18.25	$13.00 – $14.50
TVA	$15.00 – $17.25	- N.A.-

Figure 7-2

CHANGE THE ENERGY PRICE COMPONENTS

Take a look at your current rate structure. There may be certain components that are very restrictive for the way that you are doing business. The demand charge may be hurting your opportunities to optimize production. Try shifting more of the cost to the hourly kWh charge. The Time of Day rate that once was suitable may now be harmful to you in your present method of operation. A more balanced rate might also be more acceptable to your utility company.

CHOOSE NEGOTIATING TERMS BEFORE BEGINNING

Carefully evaluate all rate alternatives and choose the one best suited for your operation. Use it as your opening negotiating position. Do not let the utility company set the playing field for the negotiations.

BE SURE TO HAVE WHEELING RIGHTS

Even though retail wheeling may not be available in your area, you should include provisions for your utility company to wheel power to you from an outside source. FERC Order 888 says that they must charge you no more for this service than they charge themselves internally. This prevents them from hitting you with an unrealistically high fee for wheeling services.

SUMMARY

PREPARE, PREPARE, PREPARE. Know *your* situation and *their* situation much better than they do. This hard work <u>will</u> pay off!

Chapter 8
Power Marketers and Brokers

EXPANDING THE CLUB OF POWER MOVERS

For more than 60 years, electricity has been supplied to customers primarily by a relatively small group of regulated utility companies, each of which were given the absolute, sole right to serve customers in a well-defined territory. Rates for this service were developed in a well-choreographed regulatory process that resulted in a rate of return tied primarily to the utility's cost of service. As a result of this industry structure and regulatory process, little concern was generally placed on reducing costs or being competitive. It is clear that many generating facilities were built with weak justification, although utilities still expect to be compensated for their investment in those plants.

Today, there are major new players in the utility business who are seeking to take advantage of recent legislative and regulatory initiatives to deregulate the industry and create a competitive market. These new players include marketers and brokers who might not own any generation, but may buy and sell large quantities of power on a "commodity" basis.

NEW PRODUCTS AND SERVICES

These new players are providing important services for the deregulated power marketplace of the future. Some of these include:

- Price discovery
- Access to a broad cross section of the energy market
- Greater energy skills and capabilities
- Additional competitive pricing
- Successful management risk

Each of these services provides liquidity in the power supply market. The most visible sign of this increasing market liquidity is the explosive growth of power marketers' sales since the first quarter of 1995. In recent times, marketers' sales have grown by an average rate of 100% per quarter.

Electricity marketers and brokers are beginning to provide similar services that are supplied for natural gas end-users' requirements.

- Nomination
- Confirmations tracking
- Balancing
- Load aggregation

- Transmission management
- Invoice verification
- Miscellaneous related services

In the future, it can be expected that new products and services will be offered and/or "bundled" for the benefit of end users. These may include metering and billing services, information management, peak shaving, and other such services.

MARKETERS AND BROKERS

These new types of power providers can be divided into two categories: marketers and brokers. Marketers take ownership positions in the power sales, while brokers do not. Marketers will buy for their own account and then find a buyer (or buyers). The power marketer buys its power from any of multiple sources and then adds a "spread" on top to obtain the price to the seller. The spread between their purchase price and the sales price is their profit. The magnitude of this spread can vary widely depending on market conditions, but can often be quite large. Since a power marketer is not tied to any one set of assets, as are utility companies, they can pick a composite of the lowest cost sources at any one time. The results should be lower overall costs to the end user.

On the other hand, brokers only buy power for a specified third party. They generally do not take an ownership position in the power that they handle. Instead, a pre-established fee or commission is charged. A typical range would be from a half a mill to a mill, or more, per kWh, depending on the complexity of the assignment. Brokers bring buyers and sellers together, develop market pricing and intelligence, and maintain anonymity to all parties. For example, market pricing information is available from brokers at various delivery points, along with many physical and financial products.

GROWTH OF MARKETERS

More than 300 companies have filed for, or have already received, power marketer status at the Federal Energy Regulatory Commission. Approximately half of these companies actively are involved in trading activities. The ten largest marketers accounted for about 60% of the power marketer sales. These companies include both utility company affiliates and "independent" companies that are not otherwise associated with any utility company.

Power marketers had a major share of the market in 1997. In fact, power marketers accounted for over 1.2 billion kWh in sales during 1997, which is an increase of over 400% from the previous year. Moreover, this volume of transactions exceeded those entered into by "traditional" utilities. However, the large volume also can be attributed in part to multiple "trades" on the same quantity, as the marketers may have bought and sold the same commodity several times. Utility companies are also beginning to buy power from the marketers because it increases the flexibility in the use of their generation output.

According to Resource Data International (RDI), a major utility data analysis firm, four regions of the country – Western Systems Coordinating Council (WSCC), Mid-Atlantic Area Council (MAAC), Southeast Electric Reliability Council (SERC), and Northeast Power Coordinating Council (NPCC) – recently accounted for 80% of power marketer sales. Following is a list of RDI's conclusions for the causes of this market concentration.

- Marketers perceive these regions to be lucrative markets and are concentrating on them first as deregulation occurs.

- Fuel mixes, diverse in each region, result in more volatile markets as the marginal generation switches between the different types of generation.

- Transmission systems are better able to move power in these areas between the different companies and to the customers.

- Development of these markets tends to be self-perpetuating.

CHOOSING A BROKER OR MARKETER

When deregulation permits, an end user considering the use of a power marketer or broker might consider issuing a Request for Qualifications (RFQ) or Request for Proposal (RFP) to obtain input from multiple possible sources of energy supply. A careful evaluation of the results will be important in selecting an energy source (or sources) that meets the specific needs of the user. This approach might also uncover some creative solutions that were not previously considered. End users should carefully consider the following:

1) All power supply agreements with hidden legal problems or qualifications that limit the liability of the power supplier. An expectation of firm power supply may be dashed by the fine print of the agreement.

2) Power supply that only partially meets the end user's needs. For example, the contract may provide for base load, load following and peaking requirements throughout the year that match the end user's requirements. It is realistic to have combination of power types to meet the total requirements, but each must be carefully studied to ensure that a complete package is made available.

3) Keep in mind that power supplied by the broker or marketer must still be delivered by the local transmission and distribution utility. Be sure that these services are properly priced and included in the total "package."

4) Analyze all of the billings each month to verify their accuracy and suitability. This is also an excellent opportunity to become more familiar with patterns of energy usage by the end user.

Marketers and brokers can give end users more control over their energy supply. However, without careful management, reliance on these entities can lead to greater risk of supply interruption, and the possibility of greater errors and higher financial consequences. They must be closely managed with specifically defined contracts, in addition to consistent monitoring and control. Reporting should be carefully designed to give advanced warning of developing problems.

End users should very carefully analyze the relative benefits and drawbacks of shifting to a marketer or broker. On the one hand, there are significant financial benefits to be gained by participating in the emerging competitive markets. On the other hand, the complexities of this emerging market may take most end users into uncharted waters. Complex regulatory, legal, financial, physical and commercial issues can make a seemingly good agreement become much less valuable.

BUYER'S AGENTS

Because of the complexities and uncertainty that necessarily will result from the transition to a competitive market, end users should consider working with energy service providers that act on the end user's behalf. Such entities sometimes referred to as buyer's agents, allow end users to focus on their core business activities and "outsource" their energy procurement function to an experienced company.

Buyer's agents provide a variety of services in this regard, including:

- Understanding of the applicable legislative and regulatory environments and the ability to evaluate changes and actively participate in relevant proceedings
- Understanding of the power markets as they currently exist and the changes taking place
- Negotiating with utilities and/or marketers on behalf of the end user
- Structuring, implementing and evaluating competitive bidding, where appropriate
- Buying <u>and</u> selling energy on behalf of the end user
- Actively monitoring energy prices on a "real time" basis
- The ability to quantify, manage and mitigate transaction risk
- Understanding of the various physical and financial tools available to manage such risk

Buyer's agents provide these services on behalf of the end user and should do so on an "open book" basis. They also may provide the end user with access to additional services, including energy audits, demand-side management and energy efficiency measures, and other services aimed at reducing the end user's overall cost of energy.

SECTION 3

Preparing for Successful Gas Negotiations

Chapter 9

Emerging Issues In the Gas Industry

HISTORY OF GAS DEREGULATION
Unbundling of Natural Gas Pricing

Deregulation of the gas industry brought about the need to unbundle natural gas prices. Unbundling means the separation of services into components. Service unbundling can include the following services:

- Gas Procurement
- Arranging Pipeline Transportation
- Arranging Storage
- Balancing Services
- Load Forecasting and Nominations
- Retail Distribution
- On-System Peaking
- Backup Services and Interruption Insurance
- Metering, Accounting, Billing
- Maintenance Contracts

A customer can purchase these services separately or purchase *rebundled* combinations of these services from local distribution companies (LDC's), marketers, brokers or transporters.

Value of Unbundling

To evaluate the benefits and costs of unbundling, several items need to be considered. These include:

- Prices paid by customers
- Number and types of customers that participate
- Market share of the competitors
- Reliability of each supplier

Some customers require high reliability of supply, and will pay for it. Other customers may be willing to have lower reliability in exchange for lower prices. Specific benefits are:

- Accurate price signals ensure services are closer matched to consumer's preference

- Equitable price signals result from accurate pricing
- Efficient price signals are a by-product of accurate and equitable pricing
- Accurate, equitable and efficient pricing provides more reliable information concerning customer response to the gas suppliers
- Better regulation in a competitive unbundled environment

Costs of Unbundling

If customer savings is the short and long term objective, additional costs that result from unbundling should also be considered.

- Billing and administrative costs must be segregated to identify any additional costs required by changing suppliers
- Stranded costs may increase as a result of unbundling and competition in the gas industry
- System planning and reliability for the distribution system will continue to be the distributor s responsibility
- Low-load factor customers may be more expensive to serve
- Economics of size may be lost
- Normal business risk increases

NATURAL GAS FEDERAL REGULATIONS

Many lessons may be learned by comparing the progression of the gas market from the mid- to late 1980 s to the regulatory changes of the electric market in the mid- to late 1990's.

Natural Gas Act of 1938 (NGA)

The Federal Power Commission (FPC), now defunct, was created by the Congress to regulate all *inter*state sales and transportation of natural gas and electricity. The FPC was later succeeded by the current Federal Energy Regulatory Commission (FERC). The intended goal of the FPC was to break apart the purported monopolistic hold of the individual states, or of notable importance, of key individuals from such states, in the sale of natural gas. These sales were of gas that crossed a state boundary line at any point between pro-duction, transportation, and sales, or at any moment prior to final delivery and use. It subjected natural gas pipelines to regulatory authority.

In Section 4 of the Act, all rates, contract language and terms of service offered by a pipeline company were to be presented for public viewing, or "filed with the Commission," as tariff sheets. The Commission was granted the right to offer rebates to the companies that pay those rates, where such rates are determined to be "unjust" or "discriminatory."

In Section 5 of the Act, the Commission was granted the right to determine which filings were acceptable and which were not. When the Commission orders change pursuant to Section 5, changes must be prospective (advanced notice required), rather than retroactive in nature.

After sixty years, Section 7 of the Act is still quoted. The Commission has authority for certification and abandonment of pipelines, advanced permission to purchase or build pipelines, or to offer "jurisdictional transportation." This is the source of the term *7c certificate*, and grants the right of imminent domain for pipeline development.

Phillips vs. State of Wisconsin

The Supreme Court differentiated in this lawsuit between a) sales by a producer and the production and/or (b) gathering of gas by a pipeline. It defined Phillips as a "natural gas company," an independent producer that sold gas to *inter*state pipeline companies for transportation along *inter*state pipeline systems. Therefore, sales by Phillips were NOT exempted as "production or gathering of natural gas."

The Commission claimed jurisdiction over rates of all wholesale sales of natural gas in *inter*state commerce, whether or not by a pipeline company, and whether it occurred before, during, or after transmission by an *inter*state pipeline company. The net result was the differentiation that gas sold in *inter*state commerce would be regulated, and that gas sold in *intra*state commerce would not be regulated by FERC.

Natural Gas Policy Act of 1978 (NGPA)

In the late 1970's, after oil shortages, rolling brownouts, and long lines at the gasoline stations, the Commission set specific, escalating rates for various types of "first sales" gas (gas coming onto the market for the first time). All gas that was not committed to interstate commerce on the day before the date of enactment of the NGPA fell subject to the new rules. Each gas well was assigned a gas "section" and associated price structure depending on the spud date (date first drilled) and the type of reservoir. Decontrol of natural gas prices at the wellhead had begun.

Section 311 of the NGPA created a new environment in which "new" gas could be transported to market. This was an option previously unavailable, when "new gas" was automatically sold to the pipeline company. This opened the floodgates for a burgeoning, competitive marketplace, and set the stage for the natural gas market as it exists today.

Order 380

Minimum commodity bills were justified for fixed cost recovery, equitable cost recovery and take-or-pay recovery. (Order 380 is actually much more complex than just one sentence, but the remainder is not clearly applicable to title and land issues as much as for pipeline rate design.)

Maryland Peoples Counsel vs. FERC

The traditional pipelines created Special Marketing Programs (SMP's) in order to keep their traditional end-users on their systems rather than to abandon the traditional supply routes. These SMP's permitted the sale of "spot gas" to those industrials that could prove their potential use of alternative rules to displace natural gas. In reality, these SMP's purchased gas directly from new gas wells, or at least gas newly connected to "Pipeline B" which had previously been connected to competing "Pipeline A." The Maryland People's Counsel challenged the pipeline companies on behalf of its "member companies." These companies were "captive customers," and could not take advantage of the lower sale prices for gas (rates) that were suddenly available to others, but not to them. The Court soon sent the issue back to the FERC. Issuance of Order 436 was the eventual result.

Order 436 – Open Access (October 9, 1985)

The Commission decided to "unbundle" transportation functions (moving gas) from the merchant function (buying/selling of natural gas) for interstate pipeline companies in order to create a "workably competitive market." The pipelines may apply for "blanket certificates" with the following options:

1) Transport gas on behalf of ANY party requesting transportation, and thereby bypass the cumbersome and time-consuming previous procedure of obtaining individual certificate approvals, or

2) Choose NOT to offer "non-discriminatory transportation," and in effect, to maintain the status quo as a "closed" pipeline system. Before a pipeline could "open," it had to negotiate a settlement with its sales customers to allow them to "convert" their sales entitlements from the pipeline to equivalent levels of transportation service over a five-year period.

NOTE: Many of the resulting long term transportation agreements are "rolling over" in the latter years of this decade, and are at least partially responsible for the high instance of large volume, pipeline capacity becoming available. The net result of this may temper the "bundled" price (transportation + supply + profit margin) spikes of the mid-1990's.

The FERC also amended two other parts of its regulations.

• Certified an earlier policy statement regarding buy-outs of take-or-pay contracts

• Newly established an Optional Expedited Certificate (OEC) program under Section 7 of the NGA for pipelines adding new services, facilities and operations

Order 436 was extensively challenged in the courts, especially regarding the handling of the old take-or-pay contracts, its Contract Demand (CD) features, and its authority regarding pipeline abrogation of producer contracts.

Order 500 (August 7, 1987)

Order 500 regulated the gas under the repeal of Order 436 by the courts. Order 500 granted credits to the pipeline companies for all gas volumes that moved through its pipeline system via the newly available transportation capacity. It also allowed one-for-one credits to such pipelines against its then-mounting imbalances of take-or-pay liabilities to said producer for gas which the pipelines were forbidden to purchase, but for which prior take-or-pay contracts had been executed. It established mechanisms to handle the pass-through of settlement costs resulting from the restructuring of such take-or-pay contracts into pipeline capacity throughput. Order 500 incorporated rules for pipeline Gas Inventory Charges (GIC's). The Final Rule adjusted take-or-pay crediting provisions. In a multi-year process, the take-or-pay issue was remanded to the FERC and was eventually upheld on appeal after certain changes were made.

Order 528 (November 1, 1990) and Order 528-A (January 31, 1991)

Orders 528 and 528-A "resolved" the Take-or-Pay issue created by Order 436. These orders did not differ dramatically from Order 500 except for the method by which pipelines apply recovery expenses.

Order 636 – "Unbundling" (April 8, 1992)

*Inter*state pipelines were forced to offer each service to its customers (firm transport, storage, back-up services, etc.) independent of the use of other services. This "unbundling" offered no cross-subsidization of services, and offered "blanket certificates" to permit pipelines to offer the services at market competitive rates. That solved the problem of pipelines competing with independent marketers, but forced pipeline affiliated marketers to hold separate offices. The pipeline must treat ALL marketers the same.

The Mega-NOPR (Notice of Proposed Rulemaking) addressed pre-granted abandonment and regulations. It forced all pipelines to unbundle their "gas merchant" (buy-sell) and their pipeline transportation functions. It developed a right of first refusal for ongoing transportation contracts held by the pipelines. It addressed ratepayer recovery provisions incurred by unbundling. Finally, it mandated a Straight Fixed Variable (SFV) rate design, which reduced pipeline exposure to financial risk. The SFV lumped together all fixed pipeline expenses and then recovered them from the ratepayers via the issuance of demand charges.

Order 587-A (October 1, 1997)

This order deals with transportation operations rules, and was inspired by the Gas Industries Standards Board (GISB). It establishes operating standards (specific criteria and rules) for moving gas. It covers how gas is nominated and allocated, establishes standardized language and nomination procedures (including forms used) across all pipelines. It establishes exact time frames for both daily and monthly nominations and operations. It standardizes operational terminology and technical reporting systems. There is even an OPTIONAL standardized natural gas sales/purchase contract with its own watermark, used to protect those who use it as a guaranty that all data points in the contract are predetermined. It may be amended, or have an addendum, but the base agreement MAY NOT be altered in any way.

SOURCES OF NATURAL GAS
Producers

To review the natural gas producers, it is imperative to understand the rapid-fire changes in the demographics of the production community. With the constant mergers, and ongoing consolidations, the pure producer is something of an anomaly. This is apparent when one looks back several years ago to the ten largest natural gas producers in America. **(See Figure 9-1; 10 Largest U.S. Natural Gas Producers.)** The last year in which major producers were just that, producers, was 1993. The production listed is U.S. production only, as measured in Billion Cubic Feet per Day (BCFD).

10 Largest
U.S. Natural Gas Producers

COMPANY	Volume (BCF/D)
1. Amoco	2.37
2. Chevron	2.05
3. Exxon	1.90
4. Texaco	1.78
5. Mobil	1.52
6. Shell	1.47
7. ARCO (now Vastar)	1.11
8. Unocal	1.00
9. Phillips	0.94
10. Meridian (Burlington)	0.92

Source: Amoco Energy Trading Corporation – Ranked by BCF/D, 1993.

Figure 9-1

Since 1993 the levels of complexity in the form of alliances, joint ventures, mergers, and acquisitions have complicated the image of a producer. At the beginning of 1998, only a few of the producers listed have not merged, and two of them are rumored to currently be in the stages of final negotiations.

The function of a natural gas producer is to locate, drill for, produce, and manage natural gas reserves from below the earth's surface. While natural gas may be found in any of the traditional production basins in the United States and Canada, both onshore and offshore, it may also be found in pockets wherever the non-porous rock prevents naturally-occurring gases to either escape or move elsewhere. In the Appalachian and Rocky Mountain areas of the United States, there are numerous small pockets of gas production. They may not be of the volumes, or under high pressures, as those in the Gulf or in Texas and Louisiana, but they do exist. Methane may also be found locally in a city waste disposal system. It has all the requirements: organic materials, heat, pressure, and microorganisms. Natural gas, in varying forms of heating values and quality levels, may also be found in the refuse gas, the discarded "off-gas" left over from a prior cycle of use in industry. Therefore, though there may be no major natural gas production basins near an end-user of this energy fuel, there may be alternate methods and places for obtaining it.

Many gas buyers purchase their entire gas requirements from a marketer, a reseller who purchases gas from a producer. In 1997, a statistic was published stating that thirty-six percent of the natural gas supplies of America's marketers was purchased from other marketers.

As each company tacks on a profit margin before resale, the price of the gas has logically been inflated between the time it leaves the wellhead until it arrives at the burnertip. A savvy buyer will often attempt to purchase gas directly from a producer in order to keep costs under control through the systematic elimination of "excess margins." Marketers may add value to a transaction from the services rendered, but this segment is intended to spark a buyer to investigate direct purchases from a producer. In that situation, any marginal price difference may be split between the two parties instead of paid to a third party.

There are two kinds of producers – Majors and Independents. A comparison of their characteristics is shown in **Figure 9-2**; Comparison of Producers.

Comparison Of Producers

MAJOR PRODUCERS	INDEPENDENT PRODUCERS
Large volume of production (May have 100,000 Mcf or 1 Bcf of gas per day)	Production volumes may vary from 10Mcf/day to large volumes
Vertical integration	Usually not vertically integrated
Often international	Often domestic or North American gas producers only
Often refine gas and by-products; Process gas	Possibly also a processor, but usually not

Figure 9-2

TRANSPORTATION AND PIPELINE OPERATORS

In the United States, oil was first commercially used in the state of Pennsylvania. It was used commercially in the production of steel, and as the population grew, some of that same steel was forged into pipelines, and the transmission of natural gas as a business was made possible. The need for oil continued, as the population moved westward, and the pipelines grew.

Much of the national road system began as individually owned roads. Passengers, their vehicles and cargo were charged tolls to use these early roads made of tree bark, tree planks or crushed stone. Similarly, the first pipelines were made of hollow tree trunks, capable of carrying pools of oil a short distance. Still in existence today in southern Alberta, Canada, south of the Canadian oil and gas capital in Calgary, there are pipelines made of hollowed trees, connected and held together with hemp.

Today there are over 30 pipelines that deliver over 80% of all the gas through more than one million miles of buried pipe in the United States. Additionally, there are over 11,000 miles of interstate and intrastate natural gas pipelines under construction, or just completed. These pipelines carry over 25 billion cubic feet of additional gas per day.

An example of a current project is Tennessee's Project 2000 which includes components in New England, New York City and other key markets in the northeast and southeast United States. New power generating plants under development will replace retiring nuclear generation stations in New England. Improved high efficiency gas turbine technology makes natural gas the fuel of choice for these new facilities. (See Map: **Figure 9-3, Eastern Express Project 2000.**)

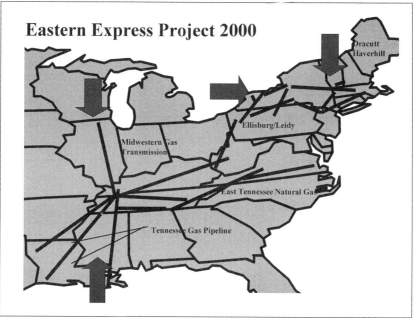

Figure 9-3

Trans-Canada Pipelines Limited plans to forge a new low cost west- to-east transportation link to move natural gas from western Canada to multiple markets in North America. The continued growth of the northeastern markets for natural gas will put a premium on the existence of a stable, efficient delivery network to get the product to market. **(See Map: Figure 9-4, TransCanada Pipe Lines Limited, and Figure 9-5; Largest Canadian Production Exporters to the United States.)**

TransCanada Pipelines Limited

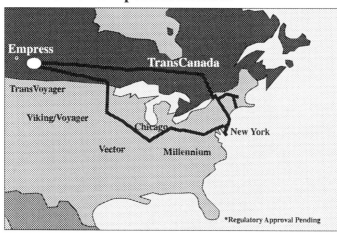

Figure 9-4

**Largest Canadian Production
Exporters To The United States**

RANK NAME	VOLUME (mmCF/D)
1. TCGS	1458
2. Pan-Alberta	1109
3. PG&E	580
4. Progas	565
5. Engage Energy	435
6. Canwest	301
7. Pan Energy	262
8. Shell	262
9. Norcen	220
10. Pan Canadian	178
11. Amoco	161
12. Husky	152
13. Poco	152
14. IGI Resources	131
15. Renaissance	107
16. SoCal	103
17. Suncor	93
18. Wascana Energy	92
19. Chevron Canada	89
20. AEC	89
Oct. 1997 – Company Names Not Updated	

Source: Brent Friedenberg Associates, Ltd., Calgary, Alberta, Canada

Figure 9-5

HUB SERVICES

Pipelines and transportation companies offer many value-added services to their customers that are possible because of the Hub System of the Pipeline Interconnections.

Parking: Provides a temporary home for stranded gas. It also provides availability, flexibility, nominations and weekend services. One might "Park" when long on gas or when favorable pricing looms just around the corner for a weather event or the first of the next month.

Wheeling: Provides short-term transportation, on either firm or interruptible basis. It also maximizes, subsidizes or substitutes use of capacity release and provides for nominations made on behalf of customer for help with mid-month balancing or excess mid-month needs.

Exchanges: Arranges trades (swaps) of gas between two (or more) willing and able parties for a mutually agreeable basis differential between the two points. It also saves dollars on transportation fees and phone bills.

Loaning: Provides a custom-tailored short-term loan of gas arranged to cover "shorts". Payback dates and prices are set prior to loan. Availability of loans depends on existing market risk and upcoming events (hurricane, holiday, etc.)

Matching: This procedure is to cover temporary imbalances – both short and long through the "match-making" of two companies with offsetting needs. Potentially eliminates hassle of "negotiating" a price. Reduces mad scrambles and minimizes losses which result from dumped gas due to looming deadlines.

Hub Transfers: This service tracks title of the gas through multiple exchanges at one transfer point. It also maintains anonymity when attempting to mask supplier identity from markets or competition. It also provides non-jurisdictional accounting functions. It is priced per title transfer. The actual function goes by many names, such as Intra-Hub Transfer (CNG-Sabine).

LOCAL DISTRIBUTION COMPANIES/MUNICIPALITIES
Local Distribution Companies (LDC's)

There are over 2,000 natural gas distribution companies, and almost 1,000 investor-owned LDC's. The ten largest investor-owned LDC's account for almost 20% of the gas delivered to customers. There are nearly 1,000 publicly owned (e.g., municipal) LDC's, marketers, brokers and aggregators. The American Public Gas Association indicates that 950 municipal, county or public utility district gas systems in the United States serve 3.8 million customers. **Ninety-five percent are served by one pipeline.**

The function of a Local Distribution Company is to:

• Serve the general public good

• Provide energy to consumers including residential, commercial, industrial, and utilities

• Provide supply, balancing, back-up services, no-notice services, and third party transportation

An LDC can be located anywhere that natural gas is used. Not every natural gas consumer utilizes the services offered by a LDC or a local/regional municipality. Many large customers are tied directly to a producer, a gathering system, an *intra*state pipeline system, or an *inter*state pipeline system.

To identify a Local Distribution Company or Municipality that provides natural gas services, refer to a copy of the:

• **Natural Gas Directory**

• **Brown's Directory (Phone: 800-598-6008)**

• **Stalsby/Wilson, "Who's Who in Natural Gas" (Phone: 301-816-8945)**

• **State Public Utility Commissions**

• **Pipeline Company Gas Control Departments**

• **Local Telephone directory "Blue Pages"**

• **Regional Natural Gas/Energy Associations**

• **Trade Organizations**

- **American Gas Association (AGA)**
- **American Society of Energy Engineers (ASEE)**
- **National Regulatory Agencies (FERC/CERI)**

Gas Suppliers: Producers/Marketers/Brokers

Many firms have been formed to sell gas as a result of federal law and regulatory policy. Customers include LDC's and industrial users. These firms provide services that are tailored to meet the needs of their clients.

Large industrial customers generally have the expertise and the financial incentive to manage their gas procurement. Smaller customers sometimes prefer to have someone else "rebundle" many of the services.

From a marketing point of view, reliability of supply will no longer be the customers' main fear or requirement. Delivery-on-demand and many new products and services will redefine the market and the transactions that flow through it. Examples of this include new strategic alliances, hedging physical commodity positions, and rebundling of non-regulated services.

The function of producers, marketers and brokers is to:

- Supply energy for consumption or resale
- Provide energy to residential, commercial, industrial and utility consumers
- Provide supply, balancing, back-up services, no-notice services
- Manage transportation (firm, interruptible, capacity released)
- Provide risk management and nomination services
- Provide volume analysis, price forecasts
- Maximize revenues, earn profits
- **(See Figure 9-6; Largest Natural Gas Marketers Ranked by Volumes.)**

Largest Natural Gas Marketers

Marketer	1996	1995
1. NCG, Chevron	Merged 1996	9.3 (merged est.)
2. Enron Capital & Trade	8.4	8.1
3. Pan Energy/Mobil	Merged 1996	7.1 (merged est.)
4. NGC	6.8	5.8
5. Amoco	5.7	5.7
6. Coral	4.2	3.7
Shell & Tejas JV; Merged with Shell;		
then Shell acquired the entire venture)		
7. Mobil (U.S.)	2.9	3.6
8. TransCanada	3.8	3.6
9. Pan Energy	4.4	3.5
10. Chevron	3.6	3.5
11. Texaco	3.2	3.2
12. Coastal	3.4	3.1
13. Utilicorp/Aquila	3.9	3.0
14. WESCO (Williams)	3.7	2.8
15. Koch	2.5	2.7
16. Exxon	2.5	2.7
17. Tenneco	2.5	2.1
18. Conoco	2.7	2.1
(Merging with El Paso)		
19. Vastar (Was ARCO) Partner w/Southern Co	2.5	2.1
20. Sonat	2.3	2.0
21. Westcoast (Merged with Coastal)	3.1	1.9
22. NorAM (Merged with Houston Industrial)	2.7	1.9
23. MidCon	2.1	1.8
24. Union Pacific	1.9	1.6
25. Western Gas	1.9	1.6
26. CNG Energy (JVs with Hydro Quebec;	1.3	1.6
Sabine; Green Mtn.)		
27. Meridian Oil (Burlington)	1.6	1.5
28. Penn Union	1.5	1.5
(Merger Pennzoil & Brooklyn Union;		
Merged with Columbia Energy Service)		

Source: Intelligence Press, Inc. (Natural Gas Intelligence, Gas, NGI's Weekly and NGI's Daily Gas Price Index. Data represents volumes reported to NGI-1995. For further information, or to order services, call (800) 427-5747. Ranked by 1995 volumes (BCF/D).

Figure 9-6

RESULTS OF COMPETITION

The downstream sectors of the natural gas industry, pipelines and LDC's have not been subjected to the forces of competition. FERC Order 636 required pipelines to provide unbundled services. In actual practice, this has not fully happened. Long-term contracts that were "grandfathered" have resulted in approximately 20% of LDC's that have been able to negotiate better transportation rates. The remaining 80% will be able to get better transportation contracts between now and the year 2005. This will bring new opportunities for customers to get lower rates.

Gas Costs

Wellhead prices for producers have changed, from below $.50 in the early 1970's, to a high point in 1982, 1983 and 1984 of near $3.50. As previously mentioned, the Natural Gas Policy Act became effective in 1985 with Orders 380 and 436. FERC Order 500 was effective in 1987, and the Natural Gas Wellhead Decontrol Act went into effect in 1989. The net results of the above changes are displayed in the following line graph. **(See Figure 9-7, Wholesale Gas Prices.)**

The costs to transport gas from the wellhead to the city gate are shown in **Figure**

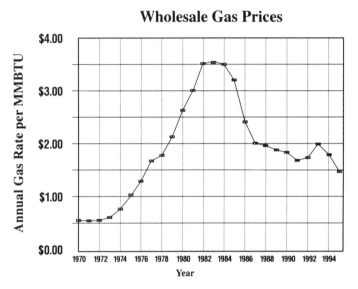

Figure 9-7

9-8, Transportation Costs. The difference between the average city gate price of approximately $3.00 per Mcf and the average wellhead or import price of approximately $2.00 per Mcf is the transportation cost of approximately $1.00 per Mcf.

The average price for gas, paid by the four major categories of users, is shown in **Figure 9-9; The Average Price of Natural Gas Delivered to Consumers in the U.S.**

Transportation Costs

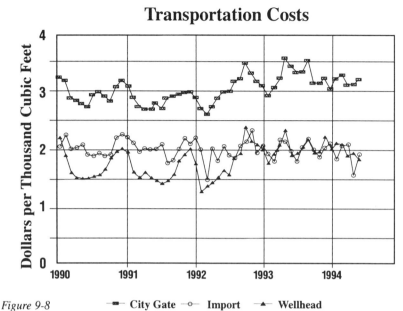

Figure 9-8   —■— **City Gate**   —○— **Import**   —▲— **Wellhead**

Average Price of Natural Gas
Delivered to Consumers

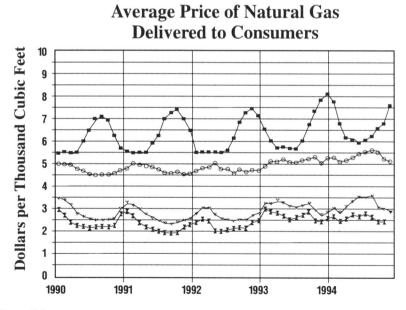

Figure 9-9   —■— **Residential**   —○— **Commercial**   —×— **Industrial**   —I— **Electric Utilities**

1) Residential pays the highest price, currently above $6.00 per mcf

2) Commercial is slightly lower – just above $5.00 per mcf

3) Industrial currently around $3.00 per mcf

4) Electric Utilities averaging around $2.50 per mcf

The variation in prices is caused primarily by the extra costs associated with extra distribution lines and customer service costs. Allocation of the consumption of natural gas in the United States for the 12 months ending September 1996 is shown in **Figure 9-10; Comparative Analysis – Consumption by End Use Sector for 12 Months.**

Comparative Analysis
Consumption by End-Use Sector
12 Months Ended September 1996 – United States

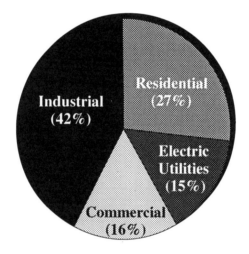

Figure 9-10

1) ***Residential***   *27%*

2) ***Commercial***   *16%*

3) ***Industrial***   *42%*

4) ***Electric Utilities***   *15%*

Declines in the cost of electricity, combined with efforts by government to instill competition in the electric utility industry, have put pressure on gas utilities to keep their costs down. Several factors affect the price of gas in today's market.

• Price of competing fuels (particularly electricity)

• Greater efficiency of gas end-uses

• Gas imports from Canada

• Normal weather conditions

• Low rates of inflation

Stranded Investments

Stranded Costs – the difference between the competitive market value and the regulated book value is the value of the potential stranded asset.

For pipelines, the process of "decontracting" for bundled pipeline services (e.g., transportation, storage, balancing) was responsible for the stranded cost issue in the pipeline sector of the natural gas industry. Given the *inter*state nature of pipelines, the stranded cost issue was handled by the FERC. In the future of retail unbundling and competition, LDC's may be faced with stranded costs associated with their distribution pipelines, storage facilities, gas supply contracts and other aspects of their systems. Unlike the electric utility industry, where several observers have offered their views on the potential extent of stranded costs, there have not been similar studies to estimate the magnitude of stranded costs for gas LDC's.

Chapter 10

Preparing for Successful
Natural Gas Negotiations

To successfully negotiate any gas contract, it is essential to understand the units of measure in which such negotiations will be most beneficial to the buyer. The most commonly used units for smaller buyers in North America are often Mcf, Ccf or therms. The unit delineation for large buyers has historically been in their energy equivalent units MMBtu, Dekatherm, or Gigajoule (Canadian).

There are untold numbers of times when unit delineation is not known by the buyer, or is not communicated clearly to the seller. Each has its own consequences. Therefore, the two parties must positively identify the exact units in which a transaction is desired. Furthermore, due to financial consequences, one must know the atmospheric pressure at which the purchase is to be transacted. For purchases, transportation, and/or storage by and between Canada and the United States, it is also essential to understand the differences of the two currencies and their interrelationship in trade.

PRICING – (Fixed Prices, NYMEX, or Basis Differentials)

Although the current price for gas is volatile, there are numerous options available. Each company has different requirements. Over the course of time, it is possible to watch the market and to identify pricing trends. It is also possible to develop corporate plans of action, and to set and meet reasonable goals.

The importance of paying attention to gas prices depends entirely on the risk level assumed by management, and the method used by management for comparing price performance. A producer may be able to offer the same options as a marketer, and the direct sale results in lower margins for everyone in each case.

BEATING THE BUDGET

Budget of $2.50/Year – If a package of gas becomes available at a fixed price to be delivered at less than $2.50/MMBtu), it would meet management's pricing criteria provided the transportation, delivery and associated risk factors also fall within pre-arranged parameters. This will not provide any security to the company that this is the lowest rate available on the market, and/or that this is the lowest rate that will become available on the market. It merely meets the requirement set forth of beating the budgetary allowance for natural gas. The buyer's task is to secure the lowest possible price available at the time, or to decide to wait until those criteria may be met.

Producer Budget: **Goal: Sell gas at the highest price**

Marketer Budget: **Goal: Buy and Sell gas at prices to maximize margin**
                         Goal: Buy and Sell gas at prices to maximize margin
                         while providing services

Purchaser Budget: **Goal: Buy gas at the lowest price**
                         Goal: Buy Gas at a price to permit pass-through of cost

Winter Prices Below $2.25 Winter – By carefully watching the market during
the winter of 1995-96, it was possible to lock in one-year or two-year fixed price
contracts in the $1.50's to mid-$1.60's/MMBtu three separate times during that
year, it was possible to purchase winter gas. During the winter of 1996-97, it was
possible to purchase gas around $2.00 to $2.05/MMBtu for one to two years. It
was possible to establish absolute price caps around $2.25/MMBtu for the win-
ter of 1997-98. To take advantage of these types of situations, one must watch
daily for pre-arranged price parameters, then respond immediately. This is the
easiest case to hedge, as long as contract arrangements are in place to permit
immediate action.

MARKET SENSITIVE PRICES

If the management goal is to purchase gas at a price equal to the current market
clearing price, then one might turn to a producer or supplier that offers a price
tied to a specific post price, or to the NYMEX to choose a daily "strike price."
The greatest challenge here is that the natural gas industry has turned into a group
of "order takers" rather than "deal makers," and the inexperienced among them
can see only as far into the future as the current trading screen. To many, the gas
price traded today in New York City for delivery in Louisiana or Alberta, Canada
must be the market-clearing price because the screen said so. This view is ques-
tionable. There actually are producers who watch the market on behalf of their
industrials, and offer guidance that benefits their customers. It is possible to find
marketers, albeit very few, who have provided similar assistance. Many charge
more for such advice (billed on a $/MMBtu basis) than it would cost to hire a
full-time supply person. There are now systems available that instantaneously
display minute-by-minute price changes, display three, six, and twelve month
"strips," basis differentials, and/or U.S. vs. Canadian exchange rates, and permit
audit of past price "ticks."

COVERING THE COMPETITION

When comparing gas costs with that of a competitor, the competition has the
upper hand if your management lets the purchase price float with the market and
the competition fix their prices at low cost. If on the other hand, your manage-

ment knows at what price the competition purchased its gas, and is able to under-cut that price by savvy purchasing, you may have competitive gas costs.

Options for Setting Price

Negotiated Price	Negotiate a fixed price for a certain length of time
Fixed Price	Price gas on a daily, monthly, quarterly, or annual basis
Posted Price	Tie the price to a daily or monthly posted price at a specific location
Futures Related Price	A locked-in price based on the NYMEX futures market, or other markets. Prices can be based on price closings for a particular day, on the last trading day, or on an average of the last three trading days of a contract for a specific upcoming time frame of one, two, three, or up to eighteen months

Sophisticated hedging techniques are available through traders, marketers, and/or sophisticated producer desks.

BASIS PRICING

Basis is the term used for the relative difference in price between two items. In natural gas, there can be basis between the cash price and the futures price.

(Basis Price = Cash Price – Future Price)

(Futures Price + Basis Price = Cash Price)

(Cash Price – Basis Price = Futures Price)

As an example: The Cash (Purchase) Price at Point "B" and the Futures Price at Point "A"

The NYMEX close at Henry Hub for Nov. '97 was $3.266/MMBtu.

The Posted Price at Princeton, NJ was $3.750/MMBtu. What was the basis?

(BASIS PRICE = CASH PRICE – FUTURES PRICE)

<u>$.486</u> = (NJ) <u>$3.750</u> – (NYMEX @ HH) <u>$3.266</u>

The use of basis pricing becomes a viable tool, especially in times of extreme pricing volatility. If an opportunity arises to lock in prices at a favorable discount to a posted price or a NYMEX closing, management might choose to use basis pricing to purchase gas to match or beat the market fluctuations.

For example, the Williams "Inside FERC" posted price in Kansas averages $.25/MMBtu above the NYMEX closing (+$.25 BASIS). Management wants to purchase gas at a price below the posted price; therefore, it may make sense to lock in a price based on a basis differential, if one can be guaranteed a price of at least $.25 below such price. Thus, if a basis of NYMEX plus, say, $.16/MMBtu can be negotiated, then the Williams price in Kansas will equal the sum of $.09, plus whatever the settlement price is on the NYMEX.

If the actual differential between the NYMEX price and the posted price on Williams P/L in Kansas turns out to be NYMEX +$.30/MMBtu, then the bet pays off. If the actual differential between the NYMEX price and the posted price on Williams P/L in Kansas turns out to be anything at all above $.25/MMBtu, then the trade was beneficial. If the actual differential is $.05/MMBtu, or anything below $.25, then the bet may or may not pay off.

One way to test if the offered basis differential is fair and reasonable is to compare it to the rolling twelve month average, or compare that rolling average to historical averages and projected average differentials.

The average price differential between two entities over the course of time is called "Strips." **(See Table 10-1; Strips – 12 Month Average.)**

Strips
12 Month Average

MONTH	BASIS TO NYMEX @ HENRY HUB	BASIS TO NYMEX @ SUMAS
January 1997	+.935	-2.000
February 1997	+.495	-1.867
March 1997	+.277	-1.424
April 1997	+.385	-1.499
May 1997	+.255	-1.225
June 1997	+.225	-1.255
July 1997	+2.53	-0.997
August 1997	+.301	-1.255
September 1997	+.423	-1.367
October 1997	+.565	-1.566
November 1997	+.595	-1.899
December 1997	+.821	-1.925

All Prices in MMBtu/US Dollars

Table 10-1

Chapter 11

Opportunities for Gas Negotiations

As a result of deregulation in the natural gas industry, there are currently a variety of opportunities available for industrial plants. Every plant will be unique, especially in the following three areas:

- **Geographic Location**
- **Total Energy Consumption**
- **Variable Load Requirements**

GEOGRAPHIC LOCATION

Where is the facility located for which natural gas is desired? In the case of multiple facilities, or multiple gas-fired burners within one facility, where is each located? If this information is not immediately available, then start with what limited data is known and collect the rest as soon as practical. Taken systematically, listed logically, and diagramed appropriately, information on these physical locations can facilitate the procurement of appropriate gas supplies. Combined with the knowledge of the pipeline systems (or distribution systems) that can deliver gas, one can determine all the available sources of gas.

Unlike electricity, natural gas may be physically transferred from one pipeline system to the next (at least on paper) in order to deliver gas to the burnertip. Along much of the entire Northeast Coast of the United States, the least cost gas is produced in Canada. An intermediary marketer may choose to price the gas as if it came from the Gulf Coast or mid-continent, but the physical gas commenced its journey (on paper at least) in Canada. That leads to a pricing issue which will be discussed later. As many maps of the United States end abruptly at the imaginary line drawn in the sand at the Canadian border, it is advisable to secure a pipeline system map that combines the United States and Canadian pipelines. A little known fact is that the longest single pipeline in North America is actually the TransCanada line. It starts in British Columbia, and flows eastward across the entire continent.

Start with the exact location of your plant, and then note the immediate vicinity of each facility and expand from there. Note the exact street address wherever possible. When a post office box is used for mail, it may also be necessary to note the actual street address, rural route, or county road to facilitate gas suppliers in the location of area pipelines. Be certain to include any applicable cross-streets. Note the Section, Township, and Range if applicable in the mid-continent states for a pinpoint location. Note the county and state or province.

When known, include the exact meter number, applicable pipeline pressures, and the direction of gas flowing through all pipes. What many have historically called a meter or a meter station, and which have been seen on charts and tables under the initials, "M," "MS," or "MT" have now been redefined to be universally called the "MLI" (meter locator indicator).

The movement to standardize technical terms began to permit people to speak effectively with one another. Before, they had used language only they understood on their local pipeline. As a non-technical case in point, just as persons in Wisconsin readily understand the phrase, "going to the bubbler," persons from the rest of the country might not immediately envision a person slurping water from a drinking fountain. In order to computerize all functions and commands to permit the Internet and electric data interchanges (EDI's) to zap data requests in nano-seconds, standardization of all terms, functions, and nominations procedures was mandated.

If possible, include a local map that delineates the natural gas pipeline distribution system. A map that identifies additional potential supply loads, such as neighboring commercial or industrial loads, might be incentive for a supplier to aggressively campaign the account.

There are commercially available pipeline maps, but many pipeline companies or local distribution companies will supply them upon request either for free, or at a minimal cost. **(See Table 11-1; Pipeline Company Maps.)** The National Survey Company in Vermont is well known for its intricately detailed maps of the various counties, local distribution areas, and states of the entire east coast of the United States.

Natural gas purchased and sold along with the companies that control the supply is divided by regions or pipeline systems. A region could include a production region (supply) or a distribution (sales) region. It could be limited to a particular state if the gas is designated for intrastate production and consumption, and then further delineated by particular intrastate pipeline systems within the area. The term "on-system" refers to those transactions that take place on a designated pipeline system. An "off-system" sale or purchase would refer to a transaction that transpired in relation to a pipeline system separate from the designated "on-system" pipeline.

Pipeline Company Maps

COLUMBIA GAS TRANSMISSION
Mr. A.V. Rao
Columbia Gulf Transmission Corp.
P.O. Box 683
Houston, Texas 77001-0683
Phone: (713) 267-4201

DUKE ENERGY FAMILY
Mr. Nick Guerrero
Duke Energy
P.O. Box 1642
Houston, Texas 77056
Phone: (713) 989-2110

    Panhandle Eastern P/L
    Trunkline Gas Co.
    Texas Eastern P/L
    Algonquin
    Northern Border P/L
    Main Pass Oil Gathering
    Main Pas Gas Gathering
    PanEnergy Field Services
    Maritimes and Northeast P/L
    (Proposed)

ENGAGE
The Coastal Corporation and
Westcoast Energy, Inc.
    Houston, Texas and
    Calgary, Alberta, Canada

MILLENNIUM GAS PIPELINE (Columbia)
Mr. Ross A. Rigler
Manager, Strategic Initiatives
Columbia Gas Transmission
2603 Augusta
Houston, Texas 77057-5637
Phone: (713) 267-4539

MISSISSIPPI RIVER INDUSTRIES MAP
Homesite Company
Box 627
Baton Rouge, Louisiana 70821
Phone: (504) 383-5369

NATURAL GAS PIPELINE COMPANY
(MIDCON)
Mr. Mike Wilder
MidCon Gas Services
701 East 22nd Street
Lombard, Illinois 60148

OZARK GAS TRANSMISSION SYSTEM
Ozark Gas Pipeline Corporation
Attn.: Transportation Services Rep.
1700 Pacific Avenue
Dallas, Texas 75201

PAN ALBERTA GAS LTD.
Collette M. Smithers
Pan Alberta Gas Ltd.
Suite 500, 707 8th Avenue SW
Calgary, Alberta, Canada T2P 3V3

TEJAS GAS CORPORATION (SHELL)
1301 McKinney – Suite 700
Houston, Texas 77010

VALERO (PG&E)
P.O. Box 500
San Antonio, Texas 78292-0500
Phone: (214) 987-8111

Table 11-1

TOTAL ENERGY CONSUMPTION

A savvy gas buyer may save thousands of dollars in avoided costs by properly preparing and transitioning available raw natural gas volume and associated pricing data into charts and graphs. A Request for Proposal (RFP) from a producer for natural gas sources, volumes and expenses should be presented in a clear, concise, readable and understandable format. The potential supplier may respond more quickly with an informed, detailed proposal instead of a minimal response or no response at all (a "no bid").

Monthly volume figures are found on the regular monthly invoice. They may be quoted for payment, depending on region, pipeline, and in many cases on negotiated terms, in physical units of Mcf (Thousand cubic feet), Ccf (Hundred cubic feet), Therms, or in heating units of Dth or Dk (Dekatherm), MMBtu (Million Btu), or Gj (Gigajoules). If no current invoices are available, then past invoices could be utilized, although they may not reflect any operational or efficiency changes that transpired between the date of such past invoice and the current time.

If the plant, facility or unit is new, then such estimated gas consumption would be estimated on an engineering study that preceded the purchase of the boilers, a furnace, or pipeline system connecting the unit, building, home or plant to the pipeline system. Ask the delivering pipeline company for daily volume charts for at least twelve calendar months prior to the current date to establish a usage pattern. The usage pattern is a critical determinant in the price one pays for natural gas. When the volume is relatively flat (daily usage is within +/- 10% of the average daily usage), then a supplier might provide a lower overall cost for supply. When the daily usage varies, or "swings" greatly, as with a municipality, an electric utility, or a fruit canning company that only cans fruit during harvesting season, then a supplier might either increase its offered overall price, or offer a dual or even triple pricing tier. This would cause the "baseload" gas to be priced on one system and the "swing" gas to be priced pursuant to either a modification of the baseload gas price or even an entirely separate pricing system. Therefore, it is important for a buyer to know what their swing load is, and to have that available in a simple diagram. Should that diagram include minimal volume swings, then it might be advantageous to include it in the supply proposal to encourage lower prices based on steady usage patterns. Two years of steady supply load could permit requests for eliminating or minimizing swing pricing.

The inverse is also true. Should that diagram indicate extreme volume swings, then it might be best NOT to include it in a proposal. See example of daily gas usage in **Figure 11-2; Daily Gas Consumption.**

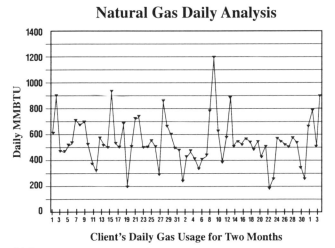

Natural Gas Daily Analysis

Client's Daily Gas Usage for Two Months

Figure 11-2

VARIABLE LOAD REQUIREMENTS

The demand requirements are variable for a plant's gas equipment. Obviously, a level load for every hour of the day throughout the year is the most desirable, especially from the producers, transporters and LDC's viewpoint. Anything less than that will cause variations in the demand, and will increase the costs for transportation significantly.

An opportunity can exist with variations in demand if the use of the gas can be interrupted from time to time, without a major loss to the process. This is normally referred to as "interruptible" or "curtailable" gas. All other gas is "firm" gas. A detailed evaluation of gas usage allows the amount required for each category to be determined.

CUSTOMER CHOICES

Following is a list of "customer choices" available under deregulated gas as sources for natural gas supply:

Gas Producer, Broker or Supplier

Many companies are available that can provide "wellhead" gas. Not all of them are actual producers. Many of them are brokers who negotiate with well owners for a block of gas from their wells. Arrangements are then made for gathering the gas and moving it from the wells to a major pipeline. Sorting through the potential sources and options can require a lot of time and effort.

The choices regarding price are basically two: 1) a fixed price; or, 2) an index price based on a published market price guideline. The decision to choose one or the other should be based on what effect price fluctuations from month-to-month, or day-to-day, will have on the cost of your products. Daily spot market cash prices tend to fluctuate widely. Monthly "hub" prices tend to be more stable than daily spot prices, and typically vary as much as $1/MMBtu.

There are many ways to hedge against wide fluctuations in gas prices. Following are some examples:

- Contract for a cap on the index price
- Contract a part of your gas at fixed prices and the balance on index
- Contract for partial firm gas and partial interruptible gas
- Contract for fixed prices with a quarterly review and update
- Contract for fixed prices and buy futures when they are low
- Contract for index prices and buy additional gas in storage

Complications may develop as you begin to evaluate the support services that are required to assure timely delivery of the gas to your plant without penalties. Review your requirements and contract for items that fit your needs such as:

- Arranging pipeline transportation

- Arranging storage
- Balancing services
- Load forecasting and nominations
- Back-up services and interruption insurance
- Responsibility for penalties

Gas Transporter or *Inter*state Gas Pipeline Company

The second contract, for transportation provided by the pipeline from the well-head to the LDC, will contain provisions for "firm", "interruptible", and other variable transportation costs.

Demand costs have always been a major part of the transportation costs. This appears to be reducing under the current market conditions.

Pipelines are afraid of losing a major part of their old standard 20 to 30 year contracts. These contracts are expected to decline to 1 to 12 years instead. This represents a negotiating opportunity for industrials that understand these facts.

Transportation costs are only part of the equation. Careful study of the reliability must be done, because of the risk of being "interrupted" during a peak demand season. Being first in line to get your gas may not assure you of reliable gas. Pipelines tend to be like airlines, and may over-book pipeline capacity. This sometimes results in the LDC or others being willing to deliver you a part or all of your gas needs during the peak season at a very high premium price. However, high prices could be better than not getting the gas at all.

Careful evaluation to understand your risks must be completed prior to signing agreements. A complete review of your alternatives and back-up positions should include:

- Purchased gas brought in from the pipeline s storage system
- Purchase permanent right-of-way on the pipeline capacity
- Purchase insurance to cover lost production or damage due to loss of gas service

Local Distribution Company (LDC)

The third contract is with the LDC for transportation from the "city gate," or the connection from the *inter*state pipeline to the LDC system. This agreement would not be necessary if you had a "bypass", or installed your own pipeline from your plant to the "*inter*state" pipeline. A pipeline map of your area can be very helpful in understanding the available options. Many industrials have installed their own line. Others have purchased the right-of-way and permits from their plant to the *inter*state pipeline and then found their LDC was willing to consider a lower transportation rate.

This agreement usually contains several options that may or may not be required. Careful study of each option should be done to assure that you understand each detail. Transportation from the "city gate" to your plant is negotiable and can vary from very low, say $.05, to very high. Most LDC's offer "backup" services in the event of "curtailment" or "interruption". Other charges could include variable fuel costs, equalization fees, administrative fees and other costs designed to insure the LDC a minimum profit. LDC charges and services can include:

- *Backup Fuel System*

  The LDC could offer other services, such as "backup" fuel, in the event of an interruption. The cost can vary depending on the amount of risk that an interruption in service represents to your plant and the understanding that the LDC has about that risk. One alternative is to install your own "backup" fuel system with propane, diesel or other alternative fuels. This would require installation of "dual fuel" burners on your plant equipment.

- *Gas Shrinkage*

  Another factor that should be considered is shrinkage. This is the amount of fuel that may be lost enroute, and discounted from your wellhead purchases by the transporter or LDC. This varies from 0% to 5% or more, and must be discussed in your negotiations.

- *Storage Alternatives*

  Another alternative to a backup fuel system is to rent space in a storage facility from the LDC or the transporter. This would provide a reserve of gas in your area to be called on if the demand on the *inter*state pipeline limited your regular supply. You would want to confirm that the LDC has capacity available under emergency conditions to deliver your gas from storage to your facility.

  Another use of storage is to purchase blocks of gas during the off-season while the prices are depressed, and use it in the peak season when prices are typically higher. Don't forget to review the cost of input and removal of the gas, as well as the cost of storage.

Gas Management Firm

Independent firms not connected with production, transportation and LDC s offer additional valuable services as well.

- *Management of Daily Nominations and Balancing*

  Tracking the actual input of gas into the delivery system from your supplier and comparing it to the actual use of gas by your plant is a management function that must be done. Your choices are to manage it yourself or negotiate an agreement with the supplier, broker, transporter or the LDC to provide that service.

The risks you face if your gas purchases do not match your consumption are many. The LDC, or the transporter, may have a "take or pay" clause in the contract. If more gas is purchased than is used in a given timeframe, you should make arrangements for balancing between what is purchased at one end of the pipe, and the amount that is consumed at the other end of the pipe. Otherwise, the LDC (or transporter) will claim title to the excess even though you paid for it. If you purchase less than is used, the LDC is generally willing to sell you the over-consumption at its regular industrial rate, which is typically higher in price than purchases from a regular supplier.

There are penalty clauses in the fine print of most gas contracts. If an outside consultant or broker is hired to manage the gas, make certain that you know who is responsible for the penalties, in the event of requirements beyond the contracted and nominated quantities.

An example of penalties is shown in the following review of one LDC's Unauthorized Use Charges:

	Past	**Current**
Supply Shortage Day	$10.00/mcf	$60.00/mcf
Supply Surplus Day	$10.00/mcf	$0.50/mcf
Non-Critical Day	$10.00/mcf	$0.50/mcf

- *Expanded Gas Use Considerations*

    One of the factors to consider is the feasibility of converting some plant equipment from electric power to gas engine drives. This has been successful for some applications, depending on the electric rates, operating requirements, and the net result of gas negotiations. If daily gas consumption is increased 50% by installation and operation of a gas engine-driven compressor, it could allow you to negotiate a lower overall gas rate. Be cautious when you consider the long-term price and availability of both the gas and the electricity.

Another consideration is to join with neighbors and form a "pool" gas purchasing arrangement. This could allow the "volume" advantage in your negotiations.

HOW TO CHOOSE A FUEL MANAGER

As the natural gas market evolves and companies are outsourcing natural gas supply, it is important to build in crosschecks on the companies that will be managing corporate assets. Following is a series of questions to present to potential fuel management groups for evaluation purposes.

Questions To Ask Potential Fuel Management Teams

1) What is their history of fuel procurement and management?

2) Who will develop and negotiate the contracts?

3) What is their qualifying past negotiating experience?

4) Will the account be assigned to an experienced account executive or placed as part of a pick-up pool?

5) How many accounts does the account executive watch? What total volume?

6) Do you know the histories of the employees?

7) Has the management group provided copies of resumes of the administrative team?

8) Have you checked the resumes and references?

9) What system do they use for contract briefing and retrieval of information?

10) What system do they use for contract briefing and retrieval of information?

11) What system do they use for retrieving futures data? Is it real time or delayed?

12) What system is used to chart data? Can they chart both real-time vs. historical data?

13) Have they historically renewed contracts in the month before the due date, within six months or a year before the due date, or after the due date?

14) What is their track record in market price forecasting?

15) Do they have economic models available? Are they flexible to fit your corporate projections?

16) Do they have any regulatory experience? Do they have access to it?

17) Can they provide dual fuel capabilities? Do they have a revenue sharing program? At what percent (10/90% vs. 50/50% vs. 90/10%)?

18) Will they be able to assist in the transition to power marketing?

19) What accountability provisions are set up to the end user?

20) What relationship does the management group have with the transporting pipeline?

21) How much capacity is owned by them, or leased by them, or controlled by them?

22) Can this company manage gas and electricity? Can they evaluate and/or improve efficiencies in plant usage of gas? Can they arrange for savings from timing of operations?

23) Does this company specialize in one region, or across the entire USA and/or Canada?

24) Does this company import/export gas across North America?

Does the Outsourcing Agreement Stipulate...

... that the work will be handled by employees of the management group?

... that a specific individual will be in charge of the account?

... that the management group will not outsource the account to others? will be able to outsource?

... that the management group will attempt to secure least cost supply? discounted capacity?

... that the management group will share revenues saved by employing alternate transport routes?

... that the person in charge of the account will actively administer the contract?

... that the management group payment shall be tied to cost savings, efficiencies, and performance?

RECOMMENDATIONS

It is not an easy task to understand and manage all the variables required to optimize the reliability and cost of gas purchases. The following list of recommended steps may help.

- Study your gas requirements and bills thoroughly.
- Evaluate all of your options and risks in detail.
- Set specific goals for your gas needs and management requirements.
- Review the impact of working with individual suppliers, transporters, local distribution companies and gas management firms compared to rebundling some or all of these services.
- Negotiate each contract on both a short-term and long-term basis. With increased competition, suppliers have even more incentive to make their clients happy.
- Make sure that you understand all the details of your new contract including expiration dates, automatic rollover dates, notice requirements for changes and *force majeure* limitations.
- Monitor the results of the pricing index that you select to assure that it will provide you the results that you have in mind.
- Cautiously shop around for the best prices, but keep in mind that it is a "small world" and price shoppers may find it increasingly hard to get a quote if you change suppliers too often.
- Compare your local LDC with other suppliers. They may have become more competitive and ready to offer more services at less cost than before.
- Keep in touch with the market changes on a regular basis.
- Assign the daily management responsibility to a staff person, or outside firm

that has demonstrated a complete understanding of the gas industry and your specific needs.

CONCLUSION

The bottom line is clear when we compare the cost of industrial gas ten years ago to the current price. The average "burner tip" price to industrials was in a range from $4.50 to $7.00 per Mcf in the mid-1980's. The mid-1990's average price range is from $2.60 to $3.60 per Mcf. This proves that the additional management that is required is well worth the additional effort.

The gas industry continues to change and evolve. A look at the traditional manner in which the gas industry is operated is shown in **Figure 11-3, Traditional Model for the Gas Industry.** In their Five-Year Strategy Plan, the GISB identified changes recommended for a new model of the gas industry. **(See Figure 11-4, New Model for the Gas Industry.)** This illustrates the difference in the traditional system that delivered gas reliably, and the new model that is designed to create more value for the customer. It will deliver gas with reliability plus flexibility, and more goods and services. All players will have access to the end customer, and this greater openness will provide new services and products.

Traditional Model for the Gas Industry

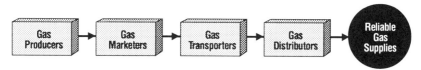

Figure 11-3

New Model for
the Gas Industry

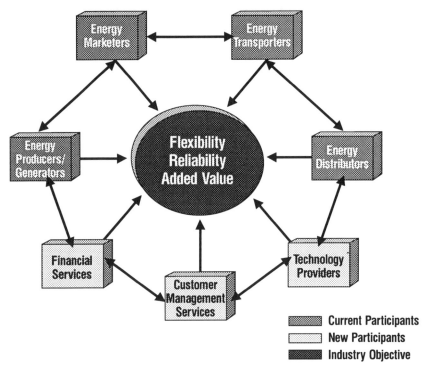

Figure 11-4

SECTION 4

Negotiating Procedures and Techniques

Chapter 12

Procedures for Successful Negotiations

Now that you have done your homework, and your team is set and ready to move into the negotiating process, the serious work begins. Following are the key elements to success in the actual negotiating process.

MANAGEMENT SUPPORT IS VITAL

The foundation for successful negotiations rests on the support of management at all levels in your organization. Many will want to continue looking to the past and the old way of doing business. Others may be suspicious of change. Still others will get impatient and expect quick results.

Get Buy-In with Extensive Communication

One of the most critical elements in successfully negotiating for a new package with your utility companies is the strong, continuing understanding and support from upper management. Most are not familiar with the involved process that is required to put together a major new agreement. Some have characterized the process as "worse than union negotiations." Often there is a fear, because deregulation is perceived to lead your company into *uncharted waters*.

The key tool is communication – early, often and in the detail necessary to give comfort on the direction of the negotiations.

Prepare Management for Long Negotiations

We Americans tend to be quite impatient. This plays into the hands of the utility companies, because they often have little to gain from moving quickly. A typical strategy is to move very slowly until the industrial company takes the position that a smaller gain is preferable to the longer wait.

Keep the "long" view. Let management know on a regular basis that the negotiations will take an extended period of time, and that long-term sacrifices should not be made for short-term gains.

Alert Management to the Likelihood of End Runs

A very real possibility is for the utility company to try to bypass the chief negotiator and go directly to upper management. This is a double loser, because you sharply diminish the authority of the chief negotiator, and eliminate a valuable negotiating tool – that of the chief negotiator to later "appeal to higher authority" (upper management).

NEGOTIATING PROCEDURES

Develop Your Negotiating Team

There are some very distinct roles to be played in the negotiating process. Do not try to combine roles, or let these players confuse their functions.

Higher Authority – Choose a senior person who normally stays remote from the negotiations. This person will be the one to whom all major decisions will be referred. This allows a particularly effective method of being able to take reinforcing or countering positions *(I thought that I could talk him into it, but he is firm on this point)* to that of the Chief Negotiator.

Chief Negotiator *(Clearly appointed)* – The Chief Negotiator (sometimes with an outside consultant) must be clearly anointed with the mantle of authority for the negotiations. Sometimes the other side will challenge this leadership, but the Higher Authority should always strongly support the Chief Negotiator.

Technical Expert – One or more separate Technical Experts are handy for reference, and to use as a justification for some of the positions taken by the Chief Negotiator. These experts could include a knowledgeable electrical engineer, a computer programmer for specialized modeling, or a financial expert.

Legal Expert – Keep the other side and your side on firm ground by referring complex legal questions to your Legal Expert. Lawyers may tend to be more confrontational than good negotiators; therefore, limited participation for this person is probably best.

Identify the Utility Company Decision-Makers and Work with Them

Take the time to observe and identify the utility company's decision-makers. When possible, work as high as possible in the organization, probably at the vice president level. Be willing to go directly to that person if you become dissatisfied with any part of the negotiations. The utility's negotiators must have decision-making ability, or you will waste your time.

Determine What They Want

There has to be some *"Candy"* in the sack for everyone. The utility companies will have some definite goals, and a number of things that they want to achieve. Draw them out, over a period of time, and incorporate their *wish list* into the negotiations as much as possible. Following are a few suggestions.

- Long-term contracts
- Not too sharp a drop in prices
- Protection from retail wheeling
- Load management
- Assistance with a subsidiary

Determine what the utility company wants, and what you want. This is the best position from which to develop a negotiating strategy. Be open with them, but do not volunteer a lot of information.

Set Common Goals Early

Try to get some consensus as to the direction of the negotiations and the boundaries set for each side. It is surprising how important this step proves to be in cutting down on delays, frustrations and recriminations later in the negotiations.

Include Some "Throw-Always"

Negotiations require *give and take* from both sides. At some time during the sessions, the other side will expect some concessions. At the beginning of the negotiation process, select some items that you would be willing to concede. Emphasize these items during discussions. At the appropriate time, give ground on them – *grudgingly*. Keep in mind that the utility company will be doing the same thing.

Go for a Bilateral Contract

In the past, all customers of utility companies have had only a unilateral rate with a small range of choices, designated from a group of tariffs approved by the state regulatory agency. Currently, the trend is toward a bilateral contract that involves both the utility company and the customer in a joint effort. This gives the customer the greatest flexibility to design a rate structure to fit his particular requirements. It is often easier to get the utility company to agree to something if they know that a precedent is not being set for all their customers.

Do Not Be Intimidated

For many years the utility companies have behaved like a big brother. They positioned themselves as "experts" in the power field and their value to the plant has been unchallenged. Some customers tend to be somewhat awed by this history and expertise. Continuation of this attitude can result in a loss of effectiveness in the negotiating process.

Do Not Be Afraid of Hurting Their Feelings

Some customers are concerned about making the utility company mad. Customers fear that they are in a vulnerable position and that the utility company may take revenge by interfering somehow with their service. Negotiations are just business with the utility, and they respect a strong (but fair) position. In addition, they have an obligation to serve you.

Do Not Be Adversarial

Conversely, the old adage, *"You can't catch flies with vinegar"* still fits in negotiating with utility companies. Many of them have been attacked in the past and are super-sensitive. They are humans, too. Listen to them and let them vent their frustrations periodically. Most people are more willing to be reasonable with someone who is *not* being unreasonable.

Go for a "Win-Win" Deal

Spend time developing the basis for a "Win-Win" deal. Determine the benefits that you can provide for the utility company. Demonstrate to them that you are not just trying to *get* something without *giving* something in return. A much better package can be negotiated with this approach. If you can get the proper terms, it just makes good sense to develop a strong partnership arrangement with your utility company.

Have Patience, but Set Deadlines the Utility Should Meet

The negotiating team must not get so impatient that major concessions are made in the name of progress. A standard negotiating technique is to delay action and refer to their "higher authority" for decisions. Be prepared for these delays.

One of the best counter tactics is to continue to set deadlines for each step of the negotiations. Continue to review the deadlines. This can subtly keep the pressure on them to meet the agreed-upon deadlines.

Be Prepared for Setbacks

Negotiating a utility contract is much like a roller coaster ride – many ups and downs before you reach your destination. It is easy to get too encouraged, and then have your hopes dashed by a setback.

Look beyond problems encountered along the way, and stay focused on long range negotiating goals that have been set. A composed negotiating posture conveys that day-to-day situations will not excessively influence goals and deadlines. A statesmanlike approach will prove to be advantageous and beneficial.

Hang Tough – Know When to Hold and When to Fold

A smooth demeanor is not the only approach to use. Controlled anger, judiciously applied, can be a powerful weapon, and may quickly bring issues back into perspective.

Similarly, be persistent. Ask the same questions multiple times. "Why can't we do it this way?" "I don't understand why you must have *that*." "Why wouldn't *this* be better?" "How would *that* provide the protection we need?"

At times, consider limited amounts of your own stonewalling to test their limits. Be a bulldog and do not give up easily. Be politely, calmly, intelligently persistent. Look for solutions. Look for reasonable compromises.

Give Something Away, but Get Something In Return

A key negotiating procedure is not to concede anything without getting something in return. Establish this precedent early in the negotiations, and it will reduce the amount of "nibbling" that the other side will do. You may also get some interesting concessions.

Keep Accurate Records of Everything

Start the negotiating process in an organized way. At every meeting, and after conversations with key players, take good notes. Be sure your notes are accurate and dated. Events may happen rapidly, and develop to a point that will cause some members of the negotiating team to have poor memories. The complexities of the discussions may overtake them. Records of meetings and discussions will provide some solid ground from which to work. It might be helpful to share meeting notes. This emphasizes the importance of consistency.

Don't Believe Everything They Say

Check important pieces of information to be sure that there are no misunderstandings, and that communications are clear. Verify and document important information, then let the other side know. This helps keep everyone aligned.

Follow-up. Follow-up. Follow-up. Leave nothing to chance. Leave nothing for someone on the other team to do improperly.

Use Your "Clubs" to "Keep them Honest"

This is a very important step, and is not as harsh as it sounds. Just insure that they play straight. It might be a regulatory position, a legal position, or something else that they want (or do not want) to concede. Do not be afraid to use these tools on carefully selected occasions.

Chapter 13

Techniques for Successful Negotiations

STUDY THE PERSONALITIES OF NEGOTIATORS

Be observant. Do not get so submerged in the negotiating process that you become oblivious and totally unaware of the human side. Some negotiators may need to be liked, some need to be disliked, or need to be needed. Despite their outward appearances, underneath it all, they have personalities like everyone else. You need to understand these human factors and the needs of different personality types. It can be an invaluable tool in successful interaction with key players in the negotiating process.

MAKE IT WORK FOR YOU

A key element in successfully working with different people is to learn to see the total person. Refrain from putting him or her into one category or the other. In most instances, you will find they are blends of several different personality types.

- Recognize and acknowledge the expertise of others.
- Look at the whole picture. You are a part of a whole. Everyone has strong points and weak points. You may be good at one thing, and weak at another. This applies to almost everyone.
- Evaluate and build on the strengths of people with whom you are negotiating. Recognize their weaknesses. Rely on the knowledge and insight you have gained, and let it work for you.

RECOGNIZE DIVERSE PERSONALITIES

Refer to **Table 13-1** for more details of different personality styles.

- *Captain* – Needs to control
- *Logical* – Needs assurance, certainty
- *Entertainer* – Needs recognition
- *Attacker* – Needs respect
- *Evader* – Needs security
- *Wanderer* – Needs freedom
- *Accommodator* – Needs to be loved and accepted
- *Attainer* – Needs and requires competency

Understanding Diverse Personalities

Style	Decision Method	Reaction to Feedback	Making It Work for You
CAPTAIN Abrupt/aloof/cold Poor delegator Confident/assertive Sense of urgency Solid eye contact Domineering/structured Perfectionists *Their strong point:* The ability to implement	Do it my way– NOW!	Turn the tables Attack	Be organized, clear, and concise; listen Show how your idea will create order, increase control and enhance results Be very direct; bottom line first Offer alternatives
LOGICAL Precise/diligent Detail oriented Poor eye contact Monotone Wants to know how/why Not creative Over-analyzes everything *Their strong point:* Analytical ability	Avoids; Re-analyzes	Ask for examples; Then they will tell you their reasons	Be factual and provide technical content and details Provide written process with all details Give them time to think it over Be specific; show them you too, respect details Be prepared – you will have to answer questions Show how each fact builds on another
ENTERTAINER Flashy Usually fun to be with Status conscious Loves center-stage Talkative/loud/jovial Self-promoters Over committed Always in a hurry *Their strong point:* Good sales and presentation skills	Sell – Sell – Sell Whatever will make self a star	Will blame others Deny any fault Rationalize	Be factual and provide technical content Focus on concept, not details Clarify facts Use storytelling to get their attention Entertain Stroke their ego
ATTACKER Hostile/angry Grouchy/argumentative Demoralizing Intimidating/nasty Looks for fault in others *Their strong point:* Ability to find loopholes	Nasty Just do it!	Interpret as personal abuse Attack	Don't take it personally; continue to show respect Be consistently responsive, not reactive Stick to the facts Prepare for lots of questions – do your homework

Table 13-1 page 1

Understanding Diverse Personalities

Style	Decision Method	Reaction to Feedback	Making It Work for You
EVADER Reserved/cautious/quiet Poor eye contact Superficial conversation Thinks paying attention to something will make it worse Fear of risk/no initiative *Their strong point:* Ability to follow instructions	Delay They love to use committees	Silence Face registers nothing	Show how your idea will insulate and protect them Try to make them feel safe, calm Relate new ideas to "known" ideas Don't expect decisiveness or initiative
WANDERER Free spirited/easy going Disorganized/impulsive Chaotic/confusing Seem distant/ambiguous Change subject often No follow-up/closure Imaginative/improvisers Hate rules/facts/structure *Their strong point:* Creative and imaginative	Who cares?	Change subject Won't listen Become distracted Give short, compact assignments Provide variety	Focus on fun; provide incentives Don't pressure or constrain Be casual, indirect, and relaxed
ACCOMMODATOR Nice/kind/agreeable Reassuring/sympathetic Give in easily Thinks of everyone else first Needs everyone to agree Hates conflict Won't confront problems Phobia about anger *Their strong point:* Great people skills	How to do "x" and keep your regard	Act hurt and cry Pretend to agree	Take a personal interest in them Show how your idea will help and please everyone Provide references and guarantees Reassure regularly Spend time with their associates, who will also be affected by your ideas
ATTAINER Peaceful/happy/serene Self-directed Fulfilled Enjoy selves and others High self-esteem Interested in others Honest/effective *Their strong point:* All of the above	Driven by effectiveness and based on research They are in charge of their own behavior Tell me more!	Thank you!	Be yourself They will adapt to you

Table 13-1 page 2

THE CONTRACT
Don t Get Impatient

At the latter end of negotiations, it is very easy to get anxious to conclude the negotiations so that you can show the great results of your efforts to the rest of the world. This is a dangerous time in the negotiating process. Be cautious, especially about items that have been postponed until the last moment.

Volunteer to Write First Draft of the Contract

Make the contract fit your needs. The first thought may be to have the utility company prepare the first drafts of all sections of the contract. This may be "easy," but may not be in your best interest. The first author has a decided opportunity to start with his version. Therefore, always volunteer to write the first draft, tailored to your advantage. The utility company will have to argue from the written text, which is harder to do. You can now "give ground" grudgingly.

Watch Their Language Very Carefully

It will be easy to agree in principle on an item and then, on careful examination, find that the text does not fit what was agreed upon. Similarly, the usual boilerplate found in most utility contracts may have some hidden "land mines" that require evasive action. Historically, their contracts have been very one-sided in their favor.

At this point, a utility lawyer can be very useful. Give him a chance to review the text of the contract, then give him specific instructions as to what his role should be in the negotiations. This allows you to get the maximum effectiveness from him at the least cost.

SUMMARY

The wheels of deregulation are swiftly turning as the federal government, state legislatures, public utility commissions and energy consumers debate issues concerning the last major monopoly power in the United States – the electric utility industry. The changes are complex and solutions may be difficult.

As the utilities cope with deregulation during the next few years, there may be some unsettling times. Questions concerning stranded costs, system management, reliability, and access will all have to be answered. However, in the end, electricity providers will more efficiently supply their customers in a competing environment. Negotiations between industrials and their suppliers will become a major force to increase profits and decrease costs.

At a recent *Energy User News* conference, Gordon Hauck from Ford Motor Company briefly summarized requirements of industrials for customer satisfaction in obtaining their power supplies. "We spend too much time and too many of our valuable resources fighting each other in the regulatory arena," he said. "Instead we need to work together to build a better future for the electricity business."

SECTION 5

**Case Study of a
Major Utility Cost Reduction Program**

Chapter 14

A *Win-Win* Industrial Experience – Case Study

Our client in the case study, a major midwestern cement plant (Plant), began a long-term plan in early 1996 to improve its profitability with a coordinated utility management program (Program). It focused strongly on negotiations with the Plant's electric utility company (Utility), in combination with in-plant enhancement of equipment, operating procedures, training, and team development. Major components of this Program are shown in **Figure 14-1, Profit Enhancement Program**. The Program has already produced significant benefits, and the savings will continue for many years into the future. The Plant has requested that its identity not be revealed.

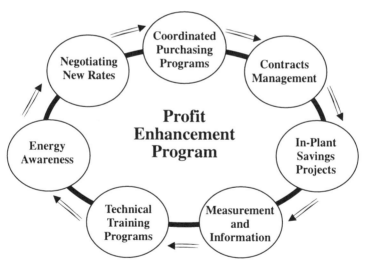

Figure 14-1

THE PROGRAM CONCEPT
The key elements of this Program are:
- Consider Utilities a Variable and Manageable Cost
- Develop a Profitable, Coordinated Approach to Managing the Utilities
- Take Advantage of the Move to Deregulation Now
- Create as Many "Win-Win" Partnerships as Possible
- Place Emphasis on Lower Cost Improvements

- Use Teamwork for Success
- Focus on Long Term, Not Quickie, Programs
- Commit Strong, Continuing Management Support
- Place Major Focus on Communications at All Levels
- Expect Production Gains

The elements of this program seem simple, but most are quite different from the historical approaches taken in many manufacturing plants. Some participants were initially quite skeptical. Old habits were deeply ingrained from many decades of repetition. Morale was poor due to recent downsizing. The future was uncertain. The challenge was great.

THE PLAN

After careful evaluation of all aspects of the Plant operation and its relationship with the Utility, the Plant management decided to begin the Program with a two-part effort. Upstream of the utility meter, we decided to push for negotiations with the Utility for a new rate package.

In the Plant, we decided to delay auditing, technical training and upgrading of the monitoring system until a communications foundation was firmly in place. The focus was on an effort to get Plant workers tuned into the concept that they could make a difference by managing their use of energy more effectively. Despite the skepticism of many key players, an Energy Team was set up, consisting primarily of hourly workers. Only two of eight members were salaried. A key member is the local president of the union. He helped to lend legitimacy to this effort.

This Team was the core of all energy activities in the Plant. More details are given in Chapter 15.

NEGOTIATING THE CONTRACT
Setting the Goal

Power costs at the Plant averaged a little less than 4 cents per kWh. We chose a target price of a little less than 3 cents per kWh to be achieved as a result of the negotiations. This was a very aggressive goal. However, at the beginning of the negotiations, we told the Utility that we wanted two and one-half cents per kWh, which allowed us some negotiating room.

"Clubs" That Were Considered

After collecting a considerable amount of information about the Plant and its electrical loads, we began the process of gathering the *"Candy"* and *"Club"* items. There was little idea about the possible outcome, but we assumed clarity would come from the preparation efforts.

One of the first alternatives considered was that of tying to a nearby municipal electric system. A relatively short distance of transmission line would be required to shift the Plant over to that system. The municipal system strongly considered the option of picking up a nice industrial load with a high load factor. In the end, however, our offer was declined. They decided that their relationship with the Utility (they purchased wholesale power from them) would be damaged severely.

The next alternative we considered was on-site generation. There was an adequate on-site gas supply, but the LDC's price was high. Evaluations included both base load generation, and peaking. No combinations of analyses were particularly attractive because our price goal was so low. In another negotiation, for another client, their beginning price for power was slightly more than 5 cents per kWh, which allowed use of on-site generation as an effective negotiating goal.

The Plant could not be shut down or relocated. In fact, the Plant made an announcement before negotiations began that the load would be expanded significantly. So much for that relocation negotiating tool!

Overall, in this particular negotiation process, the negative approach of using *"Clubs"* did not produce many attractive tools. After evaluating all the options, we focused our energies toward developing the *"Candy"* alternatives.

"Candy" That Was Considered

A major key to the successful negotiations on this project was the Utility. All involved personnel in the negotiation process were very professional, and had a clear vision of what the future held for deregulation. Their high quality personnel worked as hard as we did to reach a *Win-Win* agreement. This is in stark contrast to the backward approach experienced with other utilities.

The proposed new load turned into several attractive pieces of *"Candy"*. The prospect of a significant additional load was quite appealing. A cement plant is a very lucrative load, as shown by the **Plant's Load-Duration Curve, Figure 14-2**. Although we did not have the luxury of using the expansion as a bargaining tool before it was committed, the Utility still considered it very beneficial and worked strongly with our client on the project.

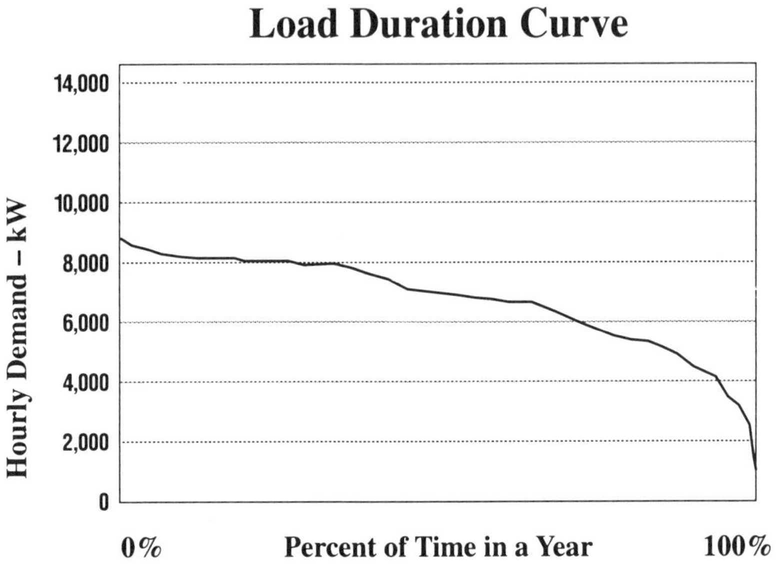

Load Duration Curve

Figure 14-2

In fact, the Utility wanted to do some electrical work pertaining to the expansion, such as the substation design and construction. They worked hard to incorporate this into the agreement and we obliged them.

The Plant had a recent bad experience with their gas supplier, during that fateful period in early February 1996 when many gas supplies were curtailed or high priced. The Utility had a gas marketing group, and we suggested the possibility of buying the Plant's gas from them - another attractive piece of *"Candy."*

Cement plants use a variety of raw materials. Under certain circumstances, they can use fly ash or bottom ash, which is a waste product from power plants. The Utility was eager to get rid of their ash, and we were able to develop an attractive offer to the Utility for part of their ash.

The Utility has an unregulated subsidiary that offers engineering services and products for sale. The stage is now set for sales of this type to the Plant.

Probably the most important piece of *"Candy"* that the Utility wanted was a longer-term contract to insure that the Plant would stay with them after deregulation. On the other hand, the Plant wanted the flexibility to take advantage of the possibility of better rates in the future resulting from deregulation. This dilemma posed a challenge for the negotiating team; but after careful planning and discussions, the issue was resolved very satisfactorily.

Other elements of the working relationship between the Utility and the Plant's parent company are confidential, as required by the state regulatory agency.

The Plant is a major user of coal, so this was another natural possibility, since utility companies are the largest users of coal. The hope was to get the advantage of the Utility Company's lower prices for coal due to greater purchasing power. After careful analyses of both firms' coal quality requirements, we realized that there were too many differences in the quality required by each company. This idea was junked.

Negotiations

After careful preparation, we began the negotiation process with the Utility. Early progress was quite encouraging and we jointly agreed to wrap up the new contract in time for a major signing ceremony in August 1996. Our actual signing was in July 1997, and the Plant started taking power under the new contract on October 1, 1997.

Our goal was for a major reduction in average power cost to the Plant. Early in the negotiations, it became clear that this would be difficult to accomplish because of the Utility Company's existing, relatively low rates.

Several years previously, we helped the Plant put an interruptible rate in place. Although only a minor amount of their load was committed to this rate, they had profited nicely from the resulting credit. However, it was not at the level that we now needed.

A cement plant runs 24 hours a day, all year, except for annual or semiannual shutdowns for maintenance. At the time negotiations began, business was good, and all Cement Plants were in a sold out condition. The Plant manager did not want to forfeit production, or suffer the potential equipment damage that an unscheduled outage might cause.

Figure 14-3, Cement Plant Process Diagram, is a simplified schematic of the elements of a cement plant operation. We investigated two possibilities. Was there a combination of surplus capacity in some of the operations in the manufacturing process, and was there adequate storage between these operations? If so, it might be possible to run some of the processes at full level, store the surplus, and shut down some of the processes during peak times.

Cement Plant
Simplified Process Diagram

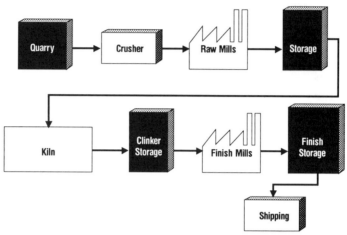

Figure 14-3

Careful analysis demonstrated that the combination of excess production capacity and storage was insufficient to be of much benefit in shaving peak loads. The savings from the interruption credit would be much less than lost production or additional maintenance costs.

This was an ideal opportunity to use a buy-though as an alternative to interrupting production. Nevertheless, its development was surprisingly difficult. The Utility was a member of a power pool that apparently did not have well-defined rules about the use of buy-throughs. There were many negotiations while resolving the terms and cost of the buy-through.

In the end, the Plant has the option of choosing the capacity that it might need to buy-through, and to make a reservation for that amount at the beginning of each year. The cost of this reservation certainly diminished the value of the buy-through; however, the result was still greatly worthwhile. It allowed the Plant to dedicate a large amount of its capacity to be interrupted – <u>without interruption.</u>

Even with this step forward, we were a long distance from reaching our price goal. Our next strategy was to evaluate a Real Time Pricing (RTP) rate, varying the price offered for power each hour. The Utility's first offer contained the use of a Customer Base Line (CBL), which is basically last year's hourly use pattern. After comparing this year's hourly use with the CBL, the customer pays the Utility when this year's usage exceeds the CBL, and receives a credit when this year's usage is less than the CBL.

A Customer Base Line is a two-edged sword. It prevents wild swings in the monthly utility bill, but it also limits the opportunity to take advantage of improvements in load management. Our computer model showed that the CBL would result in little improvement for the Plant, so we tested the model without one. The results were much better for the Plant.

There were still significant risks. See **Figure 14-4, Hourly Pricing on Summer Days,** for a plot of the hourly pricing during the summer months. Note that there are many hours when the price exceeds 20 cents per kWh. In fact, there are several times when it exceeds $1 per kWh - an astronomical price to pay for power when we are shooting for 3 cents per kWh. Anticipation of these costs was very scary to Plant management, and almost killed this strategy.

Hourly Pricing on Summer Days

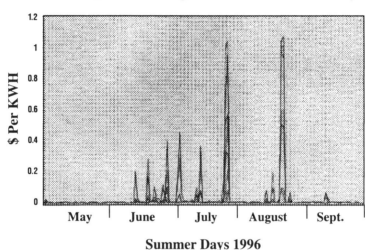

Summer Days 1996

Figure 14-4

Two things salvaged this approach. The first item was the use of the **Price-Duration Curve**, displayed in **Figure 14-5**. It shows the percentage of time each price occurs during the year. It is easy to see that most hours during the year have prices more in the realm of our goal.

The other key item was the addition of a cap on this pricing. After extensive negotiation, we were able to get the Utility to allow us to buy-through whenever the price of power reaches 4 cents per kWh. There may be occasions when the pool will have extremely high prices. However, most of the time, we can limit our hourly cost somewhat with this cap.

Price Duration Curve

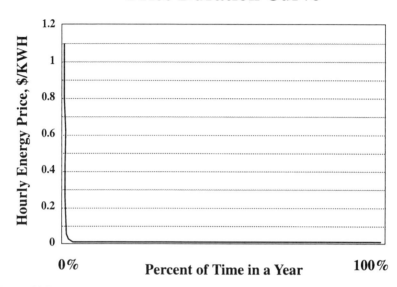

Figure 14-5

We had many of these items generally accepted by our self-imposed August 1996 deadline, and believed we were on target for execution of the contract with only several months time slippage. Then, suddenly, the Utility reorganized. The players changed. The Plant's expansion was beginning to be a significant factor, and both parties spent a significant amount of time sorting through corporate goals and setting new directions.

Everyone was dismayed and frustrated with the confusion and delays. Tempers frayed. This was a time for large amounts of patience, combined with controlled use of emotion.

Through it all, the negotiating team on both sides remained very professional. They are still good friends and have a high regard for each other.

The end results of the negotiations are very satisfying. The Plant has average power costs at their goal, and a flexible contract that will be beneficial to both them and the Utility. It is a model of the way a utility company and its customers can work together in a *Win-Win* partnership for the future.

Chapter 15

Implementing the New Rates and Plant Improvements

PLANT IMPROVEMENTS

The Plant began work on energy conservation more than six months before the new electric contract was negotiated. Management chose a broad-based effort to maximize the energy savings and a major effort has been made within the plant. This effort will continue over a period of two years.

DEVELOPMENT OF AN ENERGY TEAM

Management Buy-In

Buy-in from management was mentioned in *Chapter 12: Procedures for Successful Negotiations.* It is repeated here because of its importance. Programs cannot succeed unless one or more people who believe passionately in them are willing to take a leadership role in making them happen. Top management at the plant made a decision to establish a cross-functional Energy Team and back it strongly. Their continuing support contributed heavily to the team's success.

Plan of Action

A Plan of Action was prepared for the first Energy Team meeting. Top management's participation at this first meeting was vital as they demonstrated their complete support for the team, and introduced outside support personnel to the team. Management also provided the team with the authority and freedom to act, and encouraged individual creativity and style from all members.

Establishing the Energy Team

Our Energy Team Training Program consisted of a series of sessions that were conducted on-site. All training materials for establishing the Team, and materials for continuing evaluation exercises, were supplied for the team leader and members.

The goal was to provide training and leadership to guide the team and all plant employees in screening, identifying and implementing specific energy related projects. Key elements in the success of establishing and continuing team development are good preliminary planning, along with identifying the roles that team members and leaders will play. These were addressed in the initial meetings.

The Energy Team at the Plant consists of an excellent blend of leaders from all levels and departments, represented predominantly by hourly workers. Since the beginning, all members of the Team have been strongly motivated. This is demonstrated by the way in which they continue to pass on ideas and enthusiasm to other plant personnel. Their message is *"Everyone can make a difference!"* They continue to set an excellent example of teamwork. In addition, a review of their Mission Statement and defined functions, which they developed together, further confirms their high motivation.

- Education – Collect and disseminate energy information
- Collect Ideas – Be the central collection point for ideas on saving energy
- Evaluate Ideas – Decide which are to be implemented
- Review Ideas – Determine incentive awards
- Implement projects – Ideas are worthless unless they are implemented
- Feedback on Results – Track the results of implemented ideas

As this team evolves and assumes more responsibilities, our role is to help it stay focused, continue to clarify its vision and values, provide technical assistance, maintain good communication throughout the plant, monitor successes, recognize and reward participants, and implement identified energy saving ideas.

Team Challenges

Establishing a new team is always a very exciting project. It also requires time and patience. Even with planning and organization, there are always challenges to be met in an environment of change, new freedoms, and innovation.

- *Working Together*

  The Energy Team proved to be no different from most new teams. They brought a variety of attitudes into the group. Some were members of a team for the first time, and brought fresh ideas and an abundance of enthusiasm. Others had an attitude of been there, done that, and it probably won t work. Most of them were skeptical of management s continuing support. Almost all of them needed guidance to learn to think like managers; to learn to trust themselves; and to trust each other.

- *Training Needs*

  In addition to learning to work together and function as a team, all the members had to be trained to become "energy aware" and to recognize energy savings opportunities in the plant. They were also charged with providing this information to other plant personnel and motivating their fellow employees to become actively involved in saving energy, while saving jobs!

- *Follow the Rules*

  As the Energy Team began to realize that their collective ideas really could make a difference, their enthusiasm sometimes caused frustration. As in all plants, procedures were required to get approval for funds to implement new projects. Once projects were identified as "money and energy savers" the Team sometimes had to be reminded to "follow the rules" and fill out detailed paperwork to get projects funded and implemented properly.

- *Staying Organized*

  Time schedules and job conflicts had to be addressed early-on. The Team's accomplishments to date are highly commendable, since all members are operating in a production-oriented environment. Each member has a responsible full-time job plus being part of the Team. They continue to prioritize and set attainable goals.

ENERGY AWARENESS TRAINING

Conservation is largely a "people" item. A good Energy Awareness Program was developed and is being administered on a continuing basis to insure that utility upgrades are successful. Excellent results are being achieved by using good communication tools, supplying current information, encouraging active participation, and involving personnel at all levels in the plant.

Establishing the Need for Training

After meetings with top management and personnel at different levels in the plant, it was clear that an Energy Awareness Program would be beneficial. A plan for establishing the team was followed.

1) *Set goals*

   All staff members in each department needed a clear understanding of the impact of utilities on The Plant. They also needed to know how it related to their individual departments. An overview of the plant's utilities' history was provided and used as a benchmark in setting goals for future utility reductions. To eliminate any feelings of a threat because of change, and to eliminate misconceptions, we worked closely with all staff to assure a full awareness of the Energy Program.

2) *Prepare statistics*

   Both management and plant personnel had little knowledge or understanding of the impact of utilities or the role of energy in the plant. Simple graphs, visuals, and handouts relating to the program were used to educate at all levels.

3) *Motivate*

In conjunction with the Energy Team, personnel at all levels were encouraged to actively participate in efficiently managing utilities in the plant, and to identify savings opportunities. For example, one day was designated as Energy Awareness Day, with outside speakers from each of the utilities. Primary emphasis was placed on residential energy conservation. All personnel were invited to participate.

4) *Incentive Program*

We worked with the Human Relations staff, and discussed incentive techniques that have been used successfully in other plants. A plan was developed to establish an expanded recognition and incentive program, designed specifically for submitted energy savings ideas.

Techniques for Training

- *Set Early Goals*

It takes time for a new team to begin to function effectively. Start the team off with initial goals. The members should not be intimidated or overwhelmed, but the goals should have sufficient vision to force the team members out of their previously comfortable habits. This focus will produce cooperation, and direct the team purposefully toward greater future achievements. As a team experiences success, it develops confidence, which results in even more successes.

- *Provide Quality Speakers*

Included in all our training sessions were notebooks, written materials, forms, posters, and handouts. More important, speakers who are experts in their fields made quality presentations to the Energy Team. These sessions added variety to the program and supplied valuable information for the Team to pass on to fellow workers.

- *Involve Management*

Much of the Energy Team's success and team spirit can be attributed to the strong support of management, and their willingness to empower it with the ability to make decisions. Open communication and regular progress reviews are encouraged to maintain this high level of success.

TECHNICAL TRAINING

A training timetable was developed and training sessions scheduled for various subjects. To date, training has been given on the following selected subjects:

- Review of the Plant's Electric Bill Analysis
- Electric Motor Management/Maintenance
- Compressed Air Systems
- Pumps and Pneumatic Systems
- Air Pollution Control Systems

IN-PLANT SAVINGS PROJECTS

Identifying Projects

In order to get the maximum results, a complete plant-wide analysis was conducted. The following Plan of Action was developed from the analysis.

- Conduct a complete analysis of utility demand and energy used by all plant equipment
- Analyze a detailed audit of all utility rates and costs
- Identify specific projects that could result from changes in operating procedures, maintenance items and capital investments
- Set measurable goals that reflected the changes the Plant was willing to make

Understanding where utilities were being used, and the amounts of energy they each consumed, helped to identify those areas of greatest opportunity. We began by preparing an inventory of plant equipment, then divided the list into the following categories:

Electrical

- Determine Load: kW, horsepower, voltage, etc.
- Type of load: constant, intermittent, seasonal, etc.
  Operational hours: daily, weekly, annually
- Energy consumption: estimate in kWh

Natural Gas and Other Fuels

- Input in MMBTU
- Type of load: constant, intermittent, seasonable, etc.
  Operational hours: daily, weekly, annual
- Energy consumption: estimate in MMBtu

After the inventoried list was categorized, those pieces of equipment that had the following characteristics were identified:

- High demand peaks and low operating loads
- Excess capacity
- Inefficient system design
- Inefficient system application
- Excessive wear or high maintenance
- Waste streams

From this accumulated information, spreadsheets were developed to produce trend charts and load profiles. Additional information was gathered from in-plant sub-meter logs, operating logs, maintenance records, etc. This helped to identify billing variations and more opportunities.

Utility Rates

Research is required to understand the exact costs of utilities within a plant. We began with a copy of the new contract. From the rate information, we built a computer model that projects the cost components of the new rates. We have identified costs for the following items:

- Electric demand: kW, interruptible, real time pricing, buy through rights, other
- Electric energy: kWh, interruptible, real time pricing, buy through energy, other
- Total: cost per unit of product for all of the above components

Identifying the *cost per unit of product* is the key to producing positive bottom-line results. To obtain these results, a system was developed that is simple and reliable. All members of the Team were advised of these costs, and clearly understand their importance in dealing with any proposed utility changes in rates or contracts.

Specific Projects

From the analysis of the above information, specific projects were selected – some by the Team and some by their consultants. Changes in operating procedures, maintenance procedures and upgrades in system efficiency have been identified. Comparisons of existing equipment to industry standards or later technology will help screen these projects. The most valuable part of the analysis is to assure that each team member has an accurate understanding of the costs and what might be done to improve the system. Operators and maintenance personnel will identify additional projects from this knowledge.

Project Evaluation

A detailed estimate of the cost to implement each project required input from engineering, contractors, maintenance and suppliers. The expected savings were estimated from each project, using information from the energy use and utility rate analysis.

After an initial evaluation was completed, a group discussion with operations, production, engineering, maintenance, finance and management staff was scheduled. If you don't include everyone that is directly impacted by changes, you run the risk of overlooking an important item in your final project recommendations.

Project Implementation

Once we had approval for implementation on a project, our work had just begun. Careful planning for each step of the project helped assure "on-time and under-budget".

MEASUREMENT AND INFORMATION

A specific set of goals should be set at the beginning of any project, and a method of measuring results should be determined. In some projects this requires installation of sensors and/or connection to the data highway to assure accurate and efficient collection of data for comparison. The closer to real-time that data is gathered, compared, and translated into usable information, the more valuable it becomes.

Using the initial historical information as a base, regular comparisons are made between the "base" operating energy and other costs with the "after new rates and projects" operating energy and other costs. We do not forget to compare "before and after" unit costs of production, as well.

After each comparison, we compile the results in a simple format and share them with all team members and departments. When a real-time information technology system is available to each operator and department supervisor, the results will show up much quicker on your bottom line.

CONTRACTS MANAGEMENT

After the new contracts are in place, and the in-plant projects are completed, there are still responsibilities necessary to assure that the results will continue for many years.

- Establish a system to assure compliance with Curtailment and Interruptible requirements
- Assure Load Management Systems are producing the desired results
- Optimize savings by automating the comparison of production needs to the cost of power and fuel
- Monitor all demands, bills, power and fuel markets and energy uses
- Keep production and management informed on results

RESULTS OF PLANT TEAM

To date the team has completed six projects that continue to produce measurable results. These six projects have reduced electricity costs more than $40,000 to date with an investment of only $7,000. Other benefits include reduced maintenance and labor costs, improved safety and production improvements.

The team has seven projects that are in various stages of implementation. The estimated benefits from these projects are more than $250,000 per year in reduced electric costs with an estimated investment of no more than $150,000.

In addition to the above items, the team is developing another series of projects that is estimated to reduce the electric bill by more than $200,000 per year, and reduce maintenance costs by a minimum of $100,000 per year. This project is estimated to require an investment of $400,000.

The net result of this planned program of energy cost reduction in The Plant will be more than $580,000 *per year* reduction in operating costs. The total investment was less than $560,000. **(See Figure 15-1, Utility Management Case Study: Investments vs. Savings.)**

Utility Management Program
Investments vs Benefits

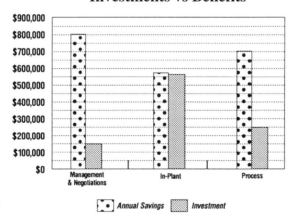

Figure 15-1

SUMMARY

Implementation of new rates and projects is not an easy task. The results will happen slowly, and will only continue with a regular effort from well-informed and supported staff members. Consider the following important items.

- Prepare, Plan, and Organize
- Set realistic and measurable goals for each identified project
- Share the information with all of your team members

The new rates and a plant-wide energy program will then be integrated into the very fabric of the Plant operation.

SECTION 6

Part 1
A Glossary of Key Electricity Terms

Access Charges – Fees charged independent producers by the owner of a transmission or distribution network.

Adequacy – Ensuring the sufficient availability of supplies in order to preserve reliability. It includes unit commitment, economic dispatch to follow load, and coordinated maintenance scheduling of individual components and planning new generation and transmission capacity.

Aggregation – the grouping of customers' electricity loads for the purpose of buying generation services. An aggregator is any entity that combines the loads of end users into a group to achieve the best rates from an electric supplier, and can include utilities, brokers, marketers, municipalities, associates and political subdivisions.

Ancillary Services: Necessary services that must be provided in the generation and delivery of electricity. As identified by the FERC, they include: coordination and scheduling services (load following, energy imbalance service, control of transmission congestion); automatic generation control (load frequency control and the economic dispatch of plants); contractual arrangements (loss compensation service); and support of system integrity and security (reactive power, or spinning and operating reserves).

APP: Affiliated Power Producer – a company (or an individual project) that generates power and is affiliated with an electric utility (e.g., Houston Industries Energy Inc. is affiliated with Houston Lighting and Power and both are subsidiaries of the holding company, Houston Industries). The differentiation between APP's and IPP's (defined below) is merely to point out that APP'S should not be allowed to make power sales to affiliated utilities due to the potential for self-dealing.

APPA: American Public Power Association – A national association representing municipally owned and other publicly owned electric utilities.

Avoided Cost – the incremental cost to an electric utility of new generation or transmission capacity (or both) which the utility avoids through conservation or purchase from another source. (For additional information, see definition in the Public Utility Regulatory Policy Act.)

Backup Power – electric energy and/or capacity supplied by an electric utility to a customer during a customer's unscheduled outage to replace energy ordinarily generated by the customer's own generation equipment. (For additional information, see definition in the Public Utility Regulatory Policy Act.)

Bilateral Model – A retail competition model that has the market organized by contracts between two parties, rather than a centralized exchange such as a power pool (see Poolco).

Broker – The company or person that matches electricity buyers and sellers. Unlike a marketer, a broker does not take title to the power.

Bulk Power Market – The market in which large amounts of electricity at high voltages is exchanged, usually from one utility to another for the purpose of resale.

Business Risk – Inherent in the operation of any business, it includes external factors such as the state of the economy, competition, product substitutes, technological obsolescence and the cost and availability of labor and raw materials.

Certificate of Need Law – This law, codified as IC. 8-1-85, requires a utility to demonstrate to the IURC the need for a new generating unit before construction begins.

Co-generation – the simultaneous production of power and useful thermal energy from the same energy source. Natural gas is burned in the turbine (essentially similar to a jet engine) which generates electricity. The hot exhaust gas is used to generate steam for process use (or used directly for heating). In a chemical plant, for example, the steam is used as heat inputs to the processes and can be used as a source of power for large rotating machinery. Steam is used in chemical reactions, separations, and purifications. Other types of Co-generation systems are possible but all of them produce both power and useful thermal energy sequentially. A Co-generation system can be a QF (defined below). (For additional information, see definition in the Public Utility Regulatory Policy Act.)

Combustion Turbine – A generating unit built to meet peak load, this is a rotary engine usually powered by natural gas.

Common Costs – Those costs incurred by a utility providing more than one service, but which cannot be assigned solely to one or another function, as they relate to the utility's overall operations.

Contract Imbalance Quantity – The cumulative difference between the quantities of a commodity received and quantities actually delivered under contract from inception through the most current billing period.

Control Area – A geographic region in which a utility is responsible for operating the power (transmission and distribution) system.

Cooperative – A business entity similar to a corporation, except that ownership is vested in members rather than stockholders, and benefits are in the form of products or services rather than profits.

Core Customers – The customers in a utility's traditional service territory. Also, called native load customers.

Corporate Unbundling – The process by which a vertically integrated firm would sell off various parts of its business to other parties. For example, an electric utility could sell of its generation, transmission, or distribution businesses.

Cost of Service Ratemaking – The traditional principles of ratemaking in which a utility's rates are set based upon the prudently incurred investment and reasonable and necessary costs to service the utility's customers plus a rate of return on the utility's investment. Properly designed cost of service ratemaking should ensure that each class of customers - residential, commercial and industrial - only pays the costs incurred to serve that class.

Cross-Subsidization – The use of resources or revenues from one part of a company's operations to fund operations in another (sometimes weaker) part of the company.

Customer Choice – the ability of an end-user (residential, commercial or an industrial customer) to purchase electricity from any supplier at negotiated rates and have that electricity delivered to a specific location.

Debt Leverage – or Debt Ratio – the proportion of borrowed capital (bonds) as compared with equity (stock), or as compared with total capital.

Demand – Rate of energy usage.

DSM: Demand-Side Management – DSM is another word for conservation programs that are meant to save money by avoiding costs for new generation. Some DSM programs are cost-effective and are proven to reduce demand, others do not work and are expensive. The utilities want customers to pay a surcharge every month for every dollar the utilities spend on DSM, even if it does not reduce demand. In addition to recovering all of their costs, they also want to charge enough to make a profit. In almost every state where there is significant application of DSM, it has resulted in higher electricity rates.

Direct Access – The ability to purchase electricity directly from the wholesale market rather than through a supplier.

Direct Assignment – A type of recovery option for stranded costs, a fee assigned to departing customers based on the costs to serve them.

Distribution – The process of transporting electric energy from a transmission system through lower voltage power lines directly to smaller end-use customers.

ECOM: Excess Cost Over Market – Those costs (typically generation costs) of a utility which are in excess of the competitive market price of electricity. ECOM is a method for calculating potential stranded investment.

Economic Dispatch – (Also Central Dispatch) The process by which a generating system consisting of multiple generating facilities is operated to maximize the efficiency of the system and minimize its operating costs. It involves using the system's most efficient (having the lowest operating costs) generating unit that is not already fully utilized when additional capacity is needed and back down or taking off the least efficient (highest operating cost) operating unit when the need for capacity is decreased.

Economic Development – Organized efforts to attract new business into an area or to encourage existing business to expand.

EDR: Economic Development Rates – Discounted rates offered to industrial customers as an incentive to locate or expand within a utility's service area and thus stimulate economic growth within that area.

Elasticity of Demand – The degree to which demand will vary with a change in price.

Embedded Costs – A cost which can no longer be avoided or minimized by the curtailment or reduction of output, because it has been incurred in some historical period of time, and cannot be varied.

EPACT: The Energy Policy Act of 1992 – EPACT clarified and expanded FERC's authority to increase competition at the wholesale level, under the appropriate safeguards. It required all utilities to provide access to their transmission system, at a reasonable price to other power producers who need to move power to a distant wholesale purchaser. Unfortunately, greater wholesale competition does not directly reduce prices at the retail level. Retail consumers are precluded from purchasing at the wholesale level. At the same time, EPACT explicitly forbade FERC from ordering retail customer choice, but EPACT preserves the right of states to take appropriate measures to do this.

Entrance Fee – Type of recovery option for stranded costs. It holds new entrants (generators) that benefit from competition responsible for a portion of transition costs.

Equity – the part of a utility owned by stockholders, represented in the financial statement of a utility as the value of outstanding common and preferred stock, retained earnings, and any additional paid-in capital.

Evergreen Contract – A contract between parties that automatically renews unless one party provides notice that it will terminate the contract after a specified time or event.

EWG: Exempt Wholesale Generator – an entity (or individual project) that generates power for wholesale sales only and is exempt from regulation under the Pubic Utility Holding Company Act of 1935 (SEC enforces this act). EWG's were created and defined by the Energy Policy Act of 1992. EWG's are a subset of NUG's (defined below). A NUG may be affiliated with an electric utility.

Exit Fee – A type of recovery option for stranded costs, this would apply to any customer switching suppliers, whether or not they continue to receive transmission service.

Externalities – this term refers to costs that are not internalized in the stated cost of a power plant. Most of these external costs are environmental (e.g., cost of additional air pollution, increased mortality due to greater emissions). Some groups want to attempt to quantify these externalities and include them in the cost of resources when deciding which resource to choose.

FERC: Federal Energy Regulatory Commission – Is responsible for regulating the price, terms, and conditions of transactions in the USA's wholesale electricity market. Also handles intrastate electricity issues.

FERC Order 888 – In April 1996, the Federal Energy Regulatory Commission (FERC) approved comprehensive rules in Order 888 which implemented nondiscriminatory, open access to transmission service for all wholesale suppliers.

Financial Risk – This is the risk arising from the method the firm uses to finance its investments and it is reflected in the firm's capital structure. It is the risk that less income will be available to the common shareholders due to the use of fixed cost financing (debt and preferred stock).

Force Majeure – A provision common in contracts that defines *force majeure* for purposes of the contract and specifies what effect *force majeure* will have on the rights and obligations of the parties under the contract. Typically, a *force majeure* clause provides that non-performance of an obligation of a party will be excused to the extent, and for so long as performance is prevented by an event of *force majeure*. *Force majeure* provisions often: exclude obligations to pay money; require the party affected to give timely notice to the other party and to use reasonable diligence to remedy the situation; and reserve for either the party NOT AFFECTED by *force majeure*, or both parties, the right to terminate or suspend the contract if the *force majeure* prevents performance for some specified period of time. [Translation: There are pre-set criteria under which one or the other party to a contract (the signatory parties) does not have to honor the exact obligations of said contract. Instead, should this pre-established event happen (explosion, government take-over, lightening, etc.), then the party having said problem could be excused from certain performance obligations and exempt from penalties. This is one of the most critical provisions to understand (potential abuse)].

Fuel Efficiency – Useful energy output divided by fuel energy input expressed as a percentage.

Functional Unbundling – A process by which a vertically integrated firm would split itself into separate functions. For example, an electric utility could organize into a generating group, a transmission group, and a distribution group. The utility would retain corporate ownership of all three business groups.

Generation – The act of converting various forms of energy input (thermal, mechanical, chemical and/or nuclear energy) into electric power. The portion of electrical service that is open to competition.

Greenfield Site – A section of land that has been set aside for industrial or commercial development, previously undeveloped. A site that was previously developed and has been set aside for renewed development or restoration is called a "brownfield site".

Grid – The transmission and distribution networks operated by electrical utilities.

Holding Company – An organization not directly engaged in the operation of any business, but which owns the stock of other companies.

Incentive Regulation – See Performance-based Ratemaking.

Incremental Cost – The additional costs incurred from the production or delivery of an additional unit of utility service, usually the minimum capacity or production that can be added. The additional cost divided by the additional capacity or output is defined as the incremental cost.

Interruptible Rates – Discounted rates that are offered to customers in exchange for the possibility of service interruptions by the utility, either at its discretion or otherwise prescribed by contract, generally in periods of high demand or short supply or during system emergencies.

Intra-Company Power Supply (Affiliate or Self-Service Wheeling) – intra-company power supply involves the transmission of excess self-generated electricity from one facility of a company to another facility of the same company. No sale is involved.

Investor Owned Utilities (IOU's) – utilities that issue common stock which is purchased by the public (as opposed to electric cooperatives or municipal utilities). These entities are taxed like other businesses, but are regulated by the state and federal governments.

IPP: Independent Power Producer – an entity (or individual project) that produces power, but does not have transmission or distribution facilities, and is not affiliated with an electric utility (e.g., Enron, Destec)

IRP: Integrated Resource Planning – the process used by utilities to determine what resources they need to meet future demand for electricity. IRP emphasizes use of conservation programs and alternative resources to meet future demand. The utilities want to use IRP as a way to get pre-approval of their decisions and as a way to automatically recover all of the costs without proper oversight.

(Note: In a truly competitive electric services industry, the need for new capacity will be controlled by supply and demand, much like other commodities.)

ISO: Independent System Operator – No exact definition yet exists, but generally it is an independent (or neutral) operator or institution that would control the regional transmission grid; that is, the transmission assets and whatever generation is needed for reliability. The ISO performs its function by controlling the dispatch of flexible plants to ensure that loads match resources available to the system. It was created as part of the implementation of the 1995 amendments to PURPA.

kVa Demand: Kilovolt-ampere Demand – The maximum number of kilovolt-ampere hours per defined time interval used by a customer. This is typically based upon the largest number of kilovolt-ampere hours used in any half-hour period of the billing period. Also, referred to as apparent power.

kW: Kilowatt Demand – The maximum number of kilowatt hours per defined time interval used by the customer. This is typically based upon the largest number of kilowatt-hours used in any half-hour period of the billing period. Also, called true power.

kWh: Kilowatt-hour – The quantity of true power multiplied by time. One thousand watts of power used for one hour equal one kilowatt-hour. Kilowatt-hours are referred to as true power because they are a measure of that portion of the kilovolt-ampere hours that can be converted from electrical energy into some other form of useful energy, such as heat, light or motion.

Load Curtailment – A temporary reduction in the amount of energy receivable by a customer due to shortfalls in supply.

Load Factor – A measure of the degree to which physical facilities, such as a power plant or gas pipeline system, are being utilized. The ratio of average output or consumption to peak output or consumption.

Load Following – The process by which a utility meets the variations in electricity demand by preparing generating units for operation under unit commitment schedules, which reflect forecasted load changes over daily, weekly, and seasonal cycles plus an allowance for random variations.

Loop Flow – A physical characteristic of a transmission system that every flow of power from a power plant to a distribution system affects the entire transmission system (or network), not just the most direct path.

Lost Revenues Approach – A method proposed by FERC to calculate stranded costs. Stranded costs are measured as the difference between revenues expected under traditional regulation and under competition.

Maintenance Power – electric energy and/or capacity supplied by an electric utility to a customer during the customer's scheduled outage to replace energy ordinarily generated by the customer's own generation equipment. For additional information, see definition of the Public Utility Regulatory Policy Act.

Marginal Cost – the cost to utilities to provide a kilowatt-hour of electricity.

Market Power – The extent to which a single firm can influence the market price.

Marketer – Any entity that buys electric energy, transmission, and other services from traditional utilities or other suppliers, and then resells those services to end-users.

MegaNOPR – Common name for the FERC NOPR (Notice of Proposed Rulemaking) that resulted in Order 888; it provides nondiscriminatory open access to the transmission system for all market participants.

Midwest ISO – An ISO attempting to be formed by more than twenty utilities in the Midwest.

Municipal Utility – An electric utility system owned and/or operated by a municipality that generates electricity and/or purchases electricity at wholesale for distribution to retail customers (residential, commercial, and industrial) which are usually, but not always, within the boundaries of the municipality.

Municipalization – The process by which a retail customer converts its legal status to that of a wholesale customer, thereby qualifying for wholesale wheeling.

NARUC: National Association of Regulatory Utility Commissions – An advisory council of governmental agencies that regulates utilities and carriers.

Natural Monopoly – An activity such as the provision of natural gas, water, and electrical service characterized by economies of scale wherein the cost of service is minimized if a single enterprise is the only seller in the market.

NERC: North American Electric Reliability Council – A nonprofit organization formed for the purpose of coordinating electric system operation and planning throughout North America. NERC was formed in 1968 in reaction to the Northeast blackout of 1965, and is organized through nine regional councils consisting of individual member electric utilities in the United States, Canada, and Mexico.

New Wholesale Contracts – Refers to contracts that will not be eligible for wholesale stranded costs at the FERC. They are defined as those contracts signed after July 11, 1994, the date the proposed rule on stranded cost recovery was published in the Federal Register.

NUG: Non-Utility Generator – the broadest term used to describe a company (or an individual project) that generates electricity but is not an electric utility (i.e., does not sell electricity on a franchise basis to retail customers).

Obligation to Serve – one of the duties of a public utility, usually referring to mandates to serve all prospective customers, to provide adequate service, and to render safe, efficiency, and nondiscriminatory service.

Order 888 – FERC's final rule on nondiscriminatory open access transmission, issued in April 1996.

Order 889 – A FERC rule establishing an open access same-time information system (OASIS) for the provision of transmission information, issued April 1996.

Pancaked Transmission Rates – Occurs when a seller attempts to wheel electricity over several control areas and pays a separate transmission charge for use of each system.

PBR: Performance-Based Ratemaking – Also referred to as "Incentive Regulation." An alternative form of ratemaking in which a utility's rates are set in a two-step process. First, rates are established on the utility's incurred and projected costs. Second, the utility is given financial incentives to reduce these costs. Properly designed PBR should pass on a portion of the cost savings to customers. However, actual experience with PBR ratemaking suggests that there is a significant possibility that rates will be set too high, and customers will not experience a reduction in prices and may experience an increase.

PoolCo – PoolCo is a highly centralized power pooling arrangement mandated by regulators that would merge the operational and marketing functions. Customer choice can only be accomplished using bilateral contracts between willing buyers and sellers. The PoolCo concept undermines customer choice by creating a new monopoly in which legitimate monopoly functions are inefficiently bundled with non-monopoly functions for the sole purpose of frustrating customer choice. In addition, the PoolCo pricing mechanism violates antitrust principles and keeps prices to most consumers artificially high by pricing all electricity at the highest clearing price.

POU – Publicly-owned utility

Power Exchange – A centralized marketplace for electricity. California has dubbed its marketplace the Power Exchange.

Power Factor (p.f.) – A ratio equal to the true power (kW) divided by apparent power (kVa).

Power Merchant – A provider of generation services, including a generator, marketer or broker, that sells generation services to end use customers at unregulated prices.

Power Pool – Two or more interconnected electric systems planned and operated to supply power in the most reliable and economical manner for their combined load requirements and maintenance programs. Essentially a regional organization of interconnected electric utilities designed to improve coordination, planning, and operation of generation and transmission facilities so as to maximize economy and reliability of service in supplying mutual energy loads. Benefits can include improved reliability of power supply, diversity of resource mix, shared spinning reserve, joint ownership of new power plants, power exchange agreements, and economy energy sales.

Price Cap Regulation – Alternative regulation plan that employs a maximum permitted rate for flexibly priced services.

PUC: Public Utilities Commission – Name of a state agency that regulates *intra*state electricity transactions and retail electric service. It must abide by FERC guidelines. Also commonly known as Public Service Commissions.

PUHCA: Public Utility Holding Company Act of 1935 – PUHCA was enacted in 1935 in response to gross financial abuses by utility holding companies. Enforcement authority was assigned to the Securities and Exchange Commission (SEC). The SEC required a structural reorganization of the holding companies that eliminated most of the problems. However, PUHCA still serves a valuable consumer protection function by creating a check against horizontal expansions through mergers and acquisitions, providing for local control and prohibiting forum shopping. Until electricity consumers have free choice of their supplier, PUHCA is needed for their protection. PUHCA prohibits acquisition of any wholesale or retail electric business through a holding company unless that business forms part of an integrated public utility system when combined with the utility's other electric business. It also restricts ownership of an electric business by non-utility corporations.

PURPA: Public Utility Regulatory Policy Act of 1978 – PURPA was enacted in 1978 to advance three goals: increased conservation of electric energy, increased efficiency in the use of facilities and resources by electric utilities, and equitable retail rates for electric consumers. One of the key elements of the Act was to require electric utilities to connect with and purchase power from qualifying facilities (QF's) at an avoided cost rate, which is equivalent to what it would have cost the utility to generate or purchase that power itself. This provision reduced the monopoly power of electric utilities by negating the utility's position as the exclusive generator of electricity.

Utilities did not, and still do not, want to purchase power from QF's, even though QF's usually produce electricity at a lower cost than do utilities. Utilities have an incentive to avoid purchasing electricity from QF's. Under historical cost-of-service rate regulation, utilities are allowed to recover all costs associated with generation of electricity plus earn a profit on their investments. When purchasing from QF's they are only allowed to charge their customers the actual cost of power purchased from QF's. Therefore, a utility's bottom line is more attractive if it can produce all of its own electricity and avoid purchasing any from QF's.

PURPA requires utilities to purchase QF power at its "avoided" cost. This means that any costs the utility would forego by not having to generate the electricity on their own existing equipment or future planned equipment should determine the price paid for the electricity generated by the QF. Each state's PUC is involved in the determination of the right avoided cost for every QF project.

Implementation of PURPA was left up to the states. State implementation has been erratic, at best. Some states required electric utilities to purchase power from QF's at prices far higher than the purchasing utility would have incurred to procure alternative sources of power. This harms consumers as well as the utilities. However, such problems were not caused by PURPA. Rather, they are the result of improper state implementation of the Act, and can and should be corrected by the state.

A very important part of PURPA for many CMA member companies is the requirement that utilities provide maintenance, backup, and supplemental power at just and reasonable, nondiscriminatory rates.

Maintenance power is supplied by the utility to the QF when a generating unit is taken out of service for scheduled maintenance. Backup power is supplied by a utility to meet a co-generator's electric requirements for those limited times when QF is unexpectedly out of service. Supplemental power is usually power that is supplied to a QF to satisfy a load in excess of the amount of power generated by the QF. Without these provisions many QF's would forego building an efficient Co-generation facility at their sites because they would have to invest extra capital to have a large reserve margin to take care of scheduled and unscheduled problems with their equipment. This reserve margin may need to be 50% to 100% compared to a normal utility reserve margin of 15% to 25%. The net effect of the loss of this provision to CMA member companies would be increased manufacturing costs, loss of competitive advantage with foreign companies, and increased emissions to the environment from the less efficient utility generation equipment.

Repeal of PURPA would strengthen the monopoly power of electric utilities by limiting the self-generation option for CMA member companies.

PX: Power Exchange – The name of the new entity in California that will establish a competitive spot market for electric power through day and hour ahead auction of generation and demand bids.

QF: Qualifying Facility – an entity (individual project) that generates power, qualifies as a co-generator or small power producer under the Public Utility Regulatory Policy Act of 1978 (PURPA), and has obtained a certificate from FERC (or self-certified). A qualifying Co-generation facility must meet certain operating and efficiency standards as specified in 18 CFR Part 292 (FERC rules implementing Section 210 of PURPA). QF's are also exempt from regulation under the Federal Power Act (FPA). QF's are NUG's, and they can be EWG's, but not necessarily. For example, an industrial that self-generates at a plant location could be a QF, but not an EWG since no power sale would be involved. QF's generally sell electricity only to utilities at avoided cost (wholesale) and current law severely limits any sales by QF's to end users (retail). (For additional information, see definition of the Public Utility Regulatory Policy Act.)

Rate Base – the accumulated capital cost of facilities purchased or installed to serve a public utility's customers, upon which the utility is allowed to earn a return. Major components include tangible and intangible plant, and working capital. Generally, rate base represents the property used and useful in public service, whose investment value is determined by a regulatory commission based on fair value, prudent investment, reproduction cost, or original cost, subject to adjustment.

Rate of Return Bandwidth Regulation – Alternative regulation plan that adjusts rates based on the earned rate of return of a utility. Also, called sliding scale, or revenue sharing.

RKVAH: Reactive kilovolt-ampere hours – The quantity of reactive power used, multiplied by time. One thousand vars of reactive power used for one hour equal one RKVAH. Reactive power is required for certain devices to operate, but it is not available as useful power on the output side of the device. Examples: motors, transformers, relays, etc.

Reactive Power – Utilized to control voltage on the transmission network, particularly, the portion of the electrical power flow incapable of performing real work or energy transfer. Reactive power is that portion of electricity that establishes and sustains the electric and magnetic fields of alternating current equipment. It must be supplied to most types of magnetic equipment, such as motors or transformers. It also must supply the reactive losses on transmission facilities. Reactive power is provided by generators, synchronous condensers, or electrostatic equipment, such as capacitors, and directly influences electric system voltage. It is measured in megavars.

Regulatory Assets – Intangibles, such as deferred debt cost and accelerated depreciation, that appear on a utility's balance sheet.

Regulatory Compact – An unwritten, unstated principle that provides for the electric utility's monopoly in a franchise area in exchange for the utility's obligation to serve all customers in that area.

Reliability – The ability to deliver uninterrupted electricity to customers on demand, and to withstand sudden disturbances such as short circuits or loss of system components, encompassing both the reliability of the generation system and the transmission and distribution system. It may be evaluated by the frequency, duration, and magnitude of any adverse effects on consumer service. The guarantee of system performance at all times and under all reasonable conditions to ensure constancy, quality, adequacy, and economy of electricity.

Reliability Council – a group of interconnected utilities in a geographical area that work cooperatively to assure system reliability. There are seven electric reliability councils in North America. In Texas, this group is known as ERCOT, and is a closed system independent of other reliability councils.

Retail – Sales covering electrical energy supplied to residential, commercial and industrial end-use purposes.

Retail Competition – (Also referred to as "Customer Choice") the ability of an end-user (residential, commercial or industrial) customer to purchase electricity from any supplier at negotiated rates and have that electricity delivered to a specific location.

Retail Wheeling – The ability of an end user to purchase electricity from a supplier of choice and transmit it over the transmission grid.

Revenue Cap – An alternative regulation plan that caps a utility's allowed revenues with an external index. Subject to the cap, a utility is permitted to maximize profits.

Revenues Lost Approach – this is the method FERC proposes to calculate stranded costs. FERC proposed this method in its Order 888. Under the "revenues lost method," the utility would calculate a customer's stranded cost liability by subtracting the current market value of power the customer currently purchases from the utility (including revenues from transmission services) from the revenues that the customer would have paid the utility had the customer continued to take service under its contract.

[Note: A "revenues lost" calculation should be discounted to take account of future cost reductions that would bring the utility's cost to market levels within a reasonable period of time. A "revenues lost" calculation should be discounted by estimates of rate discounts that customers might otherwise have received had they stayed with the incumbent utility company. A public utility should be required to demonstrate that it has taken all available measures to mitigate its stranded cost burden. These measures should be no different than the measures an unregulated business might take to adjust to a reduction in demand (sales of generation assets, corporate downsizing, etc.). Hence, a better approach to stranded costs might be to develop the value of all costs that can't be mitigated by the utility.]

ROE: Return on Equity – A measurement of the return received by common stockholders on their investment in a firm, it is the ratio of net income or earnings (after deduction of expenses) to the book value of common and preferred stock plus retained earnings. A utility is allowed (but not guaranteed) to earn the ROE authorized by a regulatory commission in a general rate case. Actual ROE is a measure of the profitability of the investment by common stockholders, and can reveal the effectiveness of management and financial decisions. Also called *Return on Common Equity.*

Rural Electric Cooperatives – organizations composed of rural areas that band together to generate or purchase electricity at wholesale and distribute the electricity to retail customers.

Senate Enrolled Act 637– Codified as IC 8-1-2.5, this statute enables the IURC to consider alternative ratemaking plans, among other things.

Service Quality – the level of performance that a utility must maintain in rendering adequate service.

Spinning Reserves – Reserve electric generating capacity that is connected to the transmission system and ready to furnish load.

Standard Offer – The rates a utility intends to charge end users under competition.

Stranded Benefits – Social programs and other regulatory assets that are currently included in utility rates that could be stranded in a competitive electricity market. (E.g.; low-income assistance, energy efficiency, research and development, and environmental mitigation.)

Stranded Costs – Revenues and assets that utilities expect to lose from a transition to a deregulated marketplace. They are currently permitted to recover these costs through their rates, but the recovery may be impeded or prevented by the advent of competition in the industry.

Stranded Investment – This is a utility term that refers to utility assets that would not be economical if a free market existed for all buyers of electricity. For example, high-cost generating facilities are generally not economical and utilities would not be able to charge the same amount for that power that they can charge as a monopoly. The utilities want someone to pay for that uneconomic investment. Determining what assets would actually be "stranded" (that is, no captive ratepayers to pay for the investment) is complicated and the utilities may not have as much uneconomic investment as they claim.

Transmission – The process of transporting electric energy in bulk on a high voltage power line from a source or sources of power supply to a point of use within a utility system or to a point of interconnection with another utility system or power grid.

Tariff – A schedule of rates, terms, and conditions for service that utilities are required to file with the state public utility commission.

TOU Rates: Time of Use Rates – Utility rates that vary depending upon the time in which services are consumed.

Transmission – That portion of a utility plant used for the purpose of transmitting electricity in bulk to other principal parts of the system or to other utility systems, or to expenses relating to the operation and maintenance of the transmission plant.

Transmission System – An interconnected group of high voltage electric lines.

Used and Useful – In service and therefore eligible for inclusion in rate base.

Voltage – The pressure or force that makes electricity flow.

Wheeling –
> **Affiliate/Self-Service Wheeling** – See "Intra-Company Power Supply."
> **Retail Wheeling** – See "Retail Wheeling and Customer Choice."
> **Wholesale Wheeling** – "Wheeling" refers to the purchase of transmission services from a utility to transport power from one place to another. Wholesale wheeling occurs when a power supplier (utility or non-utility) purchases transmission services from a utility to transport power from power plant to a wholesale customer (i.e., an investor owned utility, an electric cooperative or a municipal utility).

Wholesale – The sale of a good in large quantity for resale.

Wholesale Market – See *Bulk Power Market.*

Part 2

A Glossary of Key Gas Terms

Advanced Combined Cycle Combustion Gas Turbines – An electric generating unit that is a combustion turbine installation that uses waste heat boilers to capture exhaust energy for steam generation.

Affiliates – (See also *Subsidiaries*)

ABM's: Aggregators, Brokers and Marketers

Atmospheric Pressure – The pressure of the weight of air and water vapor on the earth's surface. The average atmospheric pressure at sea level has been defined for scientific purposes as 14.696 pounds per square inch. Various states set standards for use in measuring the volume of natural gas that is sold or processed. The American Gas Association, the FERC (Federal Energy Regulatory Commission) and all other federal agencies have adopted 14.73 pounds per square in (lbs./ in2). GISB uses 14.73 psi as the standard pressure base when no other atmospheric pressure is given. This does NOT, in any way, mandate that all gas contracts in the United States should be written with an automatic conversion to 14.73 psia.

Back-Up Service – Additional supply to assure reliability

Balancing Item – Reconciliation of actual takes versus forecasted (or nominated) use. Represents the difference between the sum of the components of natural gas supply and the sum of the components of natural gas disposition. These differences may be due to quantities lost or to the effects of data reporting problems. Reporting problems include differences due to the net result of conversions of flow data metered at varying temperature and pressure bases and converted to a standard temperature and pressure base; the effect of variations in company accounting and billing practices; differences between billing cycle and calendar period time frames; and imbalances resulting from the merger of data reporting systems which vary in scope, format, definitions, and type of respondents.

Base (Cushion) Gas – The volume of gas needed as a permanent inventory to maintain adequate underground storage reservoir pressures and deliverable rates throughout the withdrawal season. All native gas is included in the base gas volume.

BCF: Billion Cubic Feet

Btu: British Thermal Unit – A Btu is 1/180 of the heat required to raise the temperature of one pound of water from 32 degrees to 212 degrees Fahrenheit, a rise of 180 degrees. A more common definition is the amount of heat required to raise the temperature of one pound of water one degree Fahrenheit at or near 39.2 degrees Fahrenheit.

Bypass – Allowing customers to purchase from a competitor to the traditional LDC supplier. In many cases, LDCs offered discounted rates as incentive to keep customers from switching suppliers.

Bundled Service – (See Unbundled Service)

Buy/Sell Arrangement – A means of procuring gas supply where the ownership of the gas is transferred from the seller to the LDC for delivery to end-users. The LDC normally bills the buy-sell customer at its tariffed rate for system gas. The seller rebates to the customer the difference in price between the gas distributors cost of gas and the gas purchased from the seller.

Capacity – The physical capability of the facility (e.g., pipeline or power plant).

Capacity Factor – A measure of efficiency where:

$$\frac{\text{Average Load}}{\text{Rated Capacity}} \quad X \quad 100$$

Capacity Release – Selling back unneeded gas transmission capacity to the pipeline, marketer or other entities. In 1995, according to the Interstate Natural Gas Association (INGAA), about 15 percent of the gas moved on capacity that was released.

Casinghead Gas – Unprocessed natural gas produced from a reservoir containing oil; natural gas produced with oil from oil wells. Casinghead Gas contains gasoline vapors. Sometimes called "Bradenhead Gas," "Oil Well Gas," "Wet Gas," or "Solution Gas." Generally, state conservation agencies will not permit operators to produce oil wells if the associated Casinghead Gas is not sold. This gas is produced along with oil and later possibly separated into its almost pure state of methane.

City Gate – A point (measuring station) at which a LDC receives gas from a pipeline company or transmission system.

Coincidence Factor – The ratio of coincident demand to the sum of the individual demands at a specific time. Most commonly, it is the ratio of an individual (or class of) customer's demand at the time of the system peak demand. This is used in Cost-of-Service Analysis to assign costs based on a customer's contribution to the utility's peak.

Combination Utility – Typically, this refers to a utility that serves both electric and gas.

Commercial Consumption – Gas used by non-manufacturing organizations such as hotels, restaurants, retail stores, laundries and other service enterprises, and gas used by local, state, and federal agencies engaged in non-manufacturing activities.

Commodity Charge – (See also *Variable Costs*)

Consumer Price Index – A measure of aggregate prices for commodities and services typically purchased by individuals (e.g., housing, clothing, food, health care, gas, electricity, or autos). The index is generally used to gauge the change in average price levels for all commodities. By comparing the change in the price of any commodity to the change in the Consumer Price Index over a period of time, one can estimate the "real change" (i.e., the net price of general inflation in the economy) for that commodity.

Core Market – Customers that do not have competitive options and are therefore captive to a single supplier.

Correlation – (Also used as Correlation Coefficient) – A measure of the linear association between two variables, calculated as the square root of the R^2 obtained by regressing one variable against another. Correlation ranges from –1 to +1. Correlation values close to –1 or +1 show a strong correlation between the two variables (inversely proportional or directly proportional respectively). Correlation values close to zero show no correlation.

COS: Cost-of-Service – A method of allocating the costs of providing service to individual customers. Typically, there are three components of Cost-of-Service: classification, functionalization and allocation of costs. COS attempts to correlate utility costs and revenues with the service that is provided to each customer (more typically a class of customers).

Cubic Foot – A unit of volume equal to 1 cubic foot at a pressure base of 14.73 pounds standard per square inch absolute and a temperature base of 60°F.

Cubic Meter – A measurement, predominantly used in Canada, composed of the amount of natural gas that would fill an imaginary box measuring 1'x 1'x 1'. For comparison:

1 cubic meter	=	1 m³	=	35.30101 cubic feet (ft³)
1 cubic foot =		1 ft³	=	0.0283278 m³
10³m³3:		One Thousand Cubic Meters (1000 m³) of gas.		

(A box measuring 1m x 1m x 1m is a cubic meter. Fill it with natural gas.)

Customer Class – Typically, resident, commercial and industrial customers are designated as separate classes. It is not uncommon to have subclasses such as residential space heating customers. These are groups of customers with similar characteristics.

Decatherm (d, D, dt, DT, Dth, DTH) – A unit of heating value equivalent to 1,000,000 British Thermal Units. A calculation used to price gas volumes based on the amount of heat generated by a volume (Mcf) of gas. The "standard" gas unit in the United States will be the Dth pursuant to GISB under Order 587-A (1997). Therefore, to eventually convert to the dekatherm, any Local Distribution Companies, municipalities, and/or pipeline companies not yet conforming to such standard shall likely file an amended rate schedule (rate filing) detailing such transition.

Heating Value	=	Gas Volume	x	Heating Value
1 MMBtu	=	1 Mcf	x	1 Btu
1 Dth	=	1 Mcf	x	1 Btu
1 Dth	=	1,000,000 Btu		
1 Dth	=	1 MMBTU		
1 MMBTU	=	1 Dth		
1 Standard Btu	=	Btu (IT)		

Decontracting – Pipeline customers (shippers) that fail to renew their contracts for firm transportation services. The Grey Market is used to resell this unused capacity.

Deflator – An index which is used to adjust for the purchasing power of a dollar.

Demand – In economic terms, it is the inverse relationship between the price of a good and the quantity of the good that is demanded. In utility terms, it is the instantaneous (or over any specified time interval) load on the utility.

Demand Charge – (See also *Fixed Costs*) It is the amount charged to a customer (or customer class) to reflect that customer's use during a specified time interval. In Cost-of-Service analysis, the demand charge is typically based on the fixed costs associated with serving customers.

Depletion – Gas used by non-manufacturing organizations such as hotels, restaurants, retail stores, laundries, and other service enterprises, and gas used by local, State, and Federal agencies engaged in non-manufacturing activities.

Depreciation – The loss in service value not restored by current maintenance, incurred in connection with the consumption or respective retirement of a gas plant in the course of service from causes that are known to be in current operation and against which the utility is not protected by insurance; for example, wear and tear, decay, obsolescence, changes in demand and requirements of public authorities, and the exhaustion of natural resources.

Derivatives – A derivative is a financial instrument that derives its value from the value of other financial instruments or an underlying asset such as a future, forward, commodity, futures contract, stock, bond, currency, index or interest rate.

Dry Gas – Natural gas from which water content has been reduced by a dehydration process. Also natural gas containing little or no hydrocarbons commercially recoverable as liquid product. Specified small quantities of liquids are permitted by varying statutory definitions in certain sates. This is gas with the water and/or heavy hydrocarbons removed. This is commonly what is burned by a consumer.

Dry Natural Gas Production – Marketed production less extraction loss.

DSM: Demand Side Management – The planning, implementation and monitoring of utility activities designed to influence customer use of electricity in ways that will produce desired changes in a utility's load shape (i.e., changes in the time pattern and magnitude of a utility's load). Utility programs falling under the umbrella of DSM include: load management, energy efficiency, energy conservation, and innovative rates.

EBB's: Electronic Bulletin Boards (See also *GISB*) – A means of communicating the prices and availabilities of different unbundled services.

Efficiency – the ratio of inputs divided by outputs.

Electric Utility Consumption – Gas used as fuel in electric utility plants.

End-Use – Uses of energy including, but not limited to, space heating, water heating, lighting, air conditioning, refrigeration.

End-Use Load Research – Load research conducted for end-use equipment. This is done by metering these specific end uses.

Exports – Natural gas deliveries out of the continental United States and Alaska to foreign countries.

Extraction Loss – The reduction in volume of natural gas resulting from the removal of natural gas liquid constituents at natural gas processing plants.

FERC: Federal Energy Regulatory Commission – An independent agency created within the Department of Energy. The FERC is the successor of the Federal Power Commission (FPC) on September 30, 1977. The FERC is vested with broad regulatory authority over *inter*state sales of gas and electricity.

Firm Capacity (FT) – short and long-term firm

Fixed Costs – (See also *Demand Charges*)

Flared – The volume of gas burned in flares on the base site or at gas processing plants.

Gas Well Gas – Natural gas found in its own reservoir and generally not associated with any oil production.

Gathering Facility – A facility used to combine the gas from different gas wells for delivery to the pipelines.

GDP: Gross Domestic Product – Considered to be the best measure of the aggregate value of national output. GDP is equal to the Gross National Product net of residents' income from economic activity abroad (e.g., exports, repatriated profits, interest) and property held abroad minus the corresponding income of nonresident in the country (e.g., imports, profits and dividends taken out of the country).

Gigajoule – A Canadian unit of heating value equivalent to 943,213.3 Btu. The standard gas unit in Canada will be the gigajoule pursuant to GISB under Order 587-A (1997). The Gigajoule is the standard unit of natural gas heating measurement in Canada. Same as an MMBTU or a Dth, just in Canada. It is approximately 95% of 1 MMBTU or 1 Dth, DEPENDING on definition and atmospheric pressure. (NOTE: While GISB defines one standard conversion calculation, NOT EVERY metric-to-imperial measurement conversion definition will match. Therefore, there is an opportunity for both making/losing dollars.

1 Gigajoule	=	1,000,000 joules
1 Gigajoule	=	.948213 MMBTU (often rounded to .95 Dth)
1 Dekatherm	=	1.055056 gigajoules/Dth *
1 Dekatherm	=	1.054615 GJ
1 Standard Joule	=	SI system of units **

*The above GISB references – GISB Business Practice Standards. "Standards Relating to Nominations, Flowing Gas, Invoicing; Electronic Delivery Mechanisms, Capacity Release," dated 17 Oct 1996

**Published by Brent Friedenburg Associates of Calgary, Alberta, Canada

GISB: Gas Industry Standards Board – Provides for standardization of nomination practices and information pertaining to transportation services using the Internet. Order 587-Final Rule adopts 140 standards submitted by the GISB. (Standards for Business Practices of Interstate Natural Gas Pipelines, Docket No. RM96-1-000, 75 FERC 61,077 April 26, 1996). The FERC noted that much work needs to be done and set a September 30, 1996 deadline for the GISB to submit additional proposals to encompass all electronic information provided by pipelines.

GNP: Gross National Product – the total dollar value of market oriented goods and services produced by the United States economy. While the proper accounting adjustments are made, this is equivalent to adding up total income, taxes in the economy, or total sales or purchases or the total value of each industry's output.

Gray Market – (See also *Secondary Market* and *Capacity Sell Backs*)

Gross Withdrawals – Full well stream volume, including all natural gas plant liquid and non-hydrocarbon gases, but excluding lease condensate. Includes amounts delivered as royalty payments or consumed in field operations.

GSR: Gas Supply Realignment Costs) (transition costs) – An attempt to resolve the take-or-pay problem that plagued the industry during the 1980's. Under Order 636, pipelines had to realign their contracts. First, they could try to assign the contracts to former customers. The second option was to be reformed to reflect current market conditions. Of the costs incurred in reforming the contracts, 90% were allocated to firm transportation customers while 10% were allocated to interruptible customers. By June 30, 1992, pipelines agreed to absorb $3.6 billion of the estimated $10 billion in take-or-pay stranded costs. The remaining balance of $7.4 billion would be paid by consumers. Account 191 provided for direct billing of other stranded costs. The U.S. Court of Appeals for the District of Columbia July 16, 1996 held that FERC must reconsider the allocation of 10% of GSR costs to interruptible customers and explain why pipelines can pass through all of their GSR costs to customers in light of the equitable sharing procedures in Order 500 and the general cost-spreading principles in Order 636.

HDD: Heating Degree Days – A measure of how cold a location is relative to a base (normal) temperature over a period of time. The heating degree days for a single day is the difference between the base temperature and the day's average temperature. If the daily average is greater than or equal to the base, this would be a "zero" heating degree day.

Heat Content – The sum of the latent heat and sensible heat contained in a substance, above the heat contained at a selected zero condition of temperature and pressure; expressed as Btu or calories per unit of volume or weight. This measures exactly how much heat comes from the combustion of one Mcf.

Hedging – The difference between a prearranged price and the sum of: 1) the cash market city-gate price as quoted by a commodity price index, 2) a previously agreed upon retailing markup, and 3) LDC transportation charges relevant to that consumer.

Hinshaw Pipeline – A pipeline or local distribution company that has received exemption, [by Section 1(c) of the Natural Gas Act], from regulations pursuant to the Natural Gas Act. These companies transport *inter*state natural gas not subject to regulations under NGA.

Hubs – Where two or more pipelines interconnect such as the Henry Hub in Louisiana and the Moss Bluff Hub in Texas. Some practitioners use the terms "Hub" and "Market Center" interchangeably. Mechanisms to reduce the volatility of prices between various regions of the country by reducing operational and informational inefficiencies. Hubs organize trading activity at locations where prices on gas, storage, pipeline, and other services are available to all participants in the hub market, and where daily trading may be active enough to provide liquid markets.

Implicit Price Deflator – The economy's aggregate price index. Defined as the ratio of nominal GNP to real GNP.

Imports – Natural gas received in the Continental United States (including Alaska) from a foreign country.

Independent Producers – Any person who is engaged in the production or gathering of natural gas and who sells natural gas in *inter*state commerce for resale, but who is not engaged in the transportation of natural gas (other than gathering) by pipeline in *inter*state commerce.

Industrial Consumption – Natural gas used by manufacturing and mining establishments for heat, power, and chemical feedstock.

Inflation Rate – The rate of change in the economy's price level.

INGAA – Interstate Natural Gas Association of America. An association of natural gas pipelines.

Incentive Rates – (See *PBR's*)

Interruptible Capacity (IT) – According to a 1995 survey of its members by INGAA, interruptible transportation amounted to 51% in 1987 and remained fairly constant (55%, 53%, 51% and 49%) until 1992 (Order 636) when IT accounted for 42% of all gas transported. Since 1992, the amount of IT has dropped to 14% for the first half of 1995 (35% and 19% during the years 1993-1994).

***Inter*state Companies** – Natural gas pipeline companies subject to FERC| jurisdiction.

Intransit Deliveries – Redeliveries to a foreign country of foreign gas received for transportation across U.S. territory and deliveries of U.S. gas to a foreign country for transportation across its territory and redelivery to the United States.

Intransit Receipts – Receipts of foreign gas for transportation across U.S. territory and redelivery to a foreign country and redeliveries to the United States of U.S. gas transported across foreign territory.

***Intra*state Companies** – Companies not subject to FERC jurisdiction.

LDC: Local Distribution Company – The utility that is responsible for delivering gas to the customer behind the city gate (where the pipeline delivers gas to the LDC).

Lease and Plant Fuel – Natural gas used in well, field, lease operations and as fuel in natural gas processing plants.

LNG: Liquified Natural Gas – Gas that has been liquified by reducing its temperature to minus 260 degrees Fahrenheit at atmospheric pressure. Its volume is 1/600 of gas in vapor form. The carbon combinations categorized as liquified natural gas usually include methane, ethane and propane.

Load Curve – A graph that shows the shape of demand for gas (electricity) over a specified period of time (e.g., a day, month, season, year).

Load Duration Curve – A graph that shows the amount of time that gas (electric) demand is at a particular level. Demands are usually ordered from the highest to the lowest on the vertical axis. Time (e.g., a year) is on the horizontal axis.

Load Factor – The ratio, expressed as a percent (100) of the average load supplied during a designated period (e.g., hour, day) to the peak demand.

$$\frac{\text{Average Demand}}{\text{Demand}} \quad X \quad 100 \quad or \quad \frac{\text{Energy}}{\text{Peak Demand x Time}}$$

Load Research – (See also *End-Use Load Research*) Analysis of gas (or electric) usage data to better understand when and how customers use energy. This data is typically used to support load forecasting, cost-of-service and marketing programs.

Long Run – A period of time that is long enough to permit the variation of all inputs to production including capital and technological change. By way of example, long term usually describes fixed costs. Purchases of spot gas would be an example of short term costs.

LPG: Liquified Petroleum Gas – Propane, butane, or propane-butane mixtures derived from crude oil refining or natural gas fractionation. For convenience of transportation, these gases are liquified through pressure tanks.

Major *Inter*state Pipeline Company – A company whose combined sales for resale, and gas transported *inter*state or stored for a fee, exceeded 50 million thousand cubic feet in the previous year.

Marginal Cost – the change in total costs associated with a unit change in the quantity supplied. The cost of providing an additional Mcf or kWh.

Market Center – A market center has been defined as a Hub that has a pipeline, marketer or other entity identified as the operator of the interconnection. Defined as: "A market center is an area where (a) pipelines interconnect and (b) there is a reasonable potential for developing a market institution that facilitates the free interchange of gas." – FERC Order 636-B, November 27, 1992.

Marketed Production – Gross withdrawals less gas used for repressuring quantities vented and flared, and non-hydrocarbon gases removed in treating or processing operations. Includes all quantities of gas used in field and processing operations.

MCF – One Thousand Cubic Feet (1000 ft³) of gas.

1 Mcf of gas = 1.03 million Btu (Also, 1 kWh = 3.4 thousand Btu).

MMBTU – One Million British Thermal Units (1 Million Btu's). A calculation used to price gas volumes based on the amount of heat generated by a volume (Mcf) of gas. Gas in many places in the United States is purchased and sold in MMBTU quantities (Examples: Price January 1998 was $2.002/MMBTU-dry; February 1998 was $2.001/MMBTU-dry.) The term "MMBTU" is to be replaced over time (pursuant to GISB) by the term "Dekatherm." By definition, given identical heat rates, one MMBTU equals EXACTLY one MMBTU.

CAUTION: Persons new to the natural gas desk are encouraged to listen carefully for unspoken volume amounts. 1000 Mcf at a heating value of 1000 Btu is often referenced, interchangeably, as one "million Btu" or as a "thousand Btu," and just as often is referenced in single digits, thereby leaving either/both the "million" or "thousand" left unspoken. [Example: 10,000 Mcf at a heating content of 1,000 Btu (also seen as 1.000 Btu) could be requested as: "I need 10,000 per day," or "I need 10 Million per day," or simply, "I need 10 per day." Caution to avoid confusion is issued to both the buyer and the seller to translate that request into a total dollar amount prior to acceptance or rejection. At a sales price of $2.00/MMBTU, 10 MMBTU/Dth would equate to $20.00, whereas, 10,000 MMBTU/Dth would equate to $20,000.]

MMCF – Million Cubic Feet

Native Gas – Gas in place at the time that a reservoir was converted to use as an underground storage reservoir as in contrast to injected gas volumes.

Natural Gas – A naturally occurring mixture of hydrocarbon and non-hydrocarbon gases found in porous geologic formations beneath the earth's surface, often in association with petroleum. The principal constituent is methane. It exists in a gaseous form. It may be transported via a pipeline system. It may be condensed and/or refrigerated, and be transported in these forms via ship, rail, truck and/or bus.

```
            H
($CH_4$)    H    C    H      1 Carbon with 4 Hydrogen Molecules
            H
```

Natural Gas Liquids – This category is inclusive of more differentiations of natural carbons than LNG or LPG, although it includes the particular carbon molecules that could comprise both of the above. Natural gas liquids include the following carbon components: ethane (C_2H_6), propane (C_3H_8), butanes (C_4H_{10}), pentanes (C_5H_{12}), and hexanes (C_6H_{14}).

NCP: Non-Coincident Peak Demand – The sum of two or more individual demands that do not occur in the same (coincident) time interval. Mathematically, the NCP can be equal to, but is almost always greater than, the coincident peak demand.

NGPA: Natural Gas Policy Act of 1978 – Signed into law in November 1978, the NGPA is a framework for the regulation of most facets of the natural gas industry.

Explanatory Note: The "balancing item" category represents the difference between the sum of the components of natural gas supply and the sum of the components of natural gas disposition. These differences may be due to quantities lost or to the effects of data reporting problems. Reporting problems include differences due to the net result of conversions of flow data metered at varying temperatures and pressure bases and converted to a standard temperature and pressure base; the effect of variations in company accounting and billing practices; differences between billing cycles and calendar periods; and imbalances resulting from the merger of data reporting systems, which vary in scope, format, definitions, and type of respondents.

No Notice Service – Rebundled pipeline services.

Nominal – An adjective that describes any monetary magnitude measured in current rather than "real" dollars. For example, nominal Total personal income is the current dollar value of Total Personal Income through time not adjusted to reflect the general levels of price increase in the economy through time.

Non-Firm Purchase – An "as available" basis. There is no commitment to serve.

Non-Hydrocarbon Gases – Typical non-hydrocarbon gases that may be present in reservoir natural gas are carbon dioxide, helium, hydrogen sulfide, and nitrogen.

On System Sales – Sales to customers where the delivery point is a point on, or directly interconnected with, a transportation, storage, and/or distribution system operated by the reporting company.

PBR's: Performance-Based Rates – Sometimes referred to as "Incentive Rates," PBR's may take many forms including: "price caps," "yardstick regulation" (e.g., comparing a utility to other utilities), "sliding scale" regulation (i.e., where the customers and the shareholders share benefits and costs), or hybrids.

PSIA – Absolute Pressure – The factor for pressure used in determining a gas' volume, expressed in terms of pounds of pressure per square inch absolute, or pressure that the gas would exert on the walls of a one-cubic-foot container if it were at a given absolute pressure, or if one of numerous artificial stipulations were placed on the gas. One might see American gas priced at $2.40/Dth at 14.73 psia. (Translation: $2.40 per Dth at 14.73 psia. Caution: It is important financially to determine, both up front and in any future audit, whether a negotiated price is psig or psia, because there may be a financial advantage to one party or the other. It may be that an artificial atmospheric pressure is applied that will be to the advantage of the party proposing the absolute pressure. A simple way to remember atmospheric pressure is to recall that atmospheric pressure is much greater under water, and much less in outer space, and is calculated by the height above sea level of a particular location.

PSIG – Gauge Pressure – The factor for pressure used in determining a gas' volume, expressed in terms of pounds of pressure per square inch gauge, or pressure that the gas would exert on the walls of a one-cubic-foot container. This is an engineer's or gas controller's measurement. It determines the actual pressure, without any adjustments. Often artificial and/or calculated adjustments are made to a straight psig measurement that ultimately results in higher or lower total calculated delivered volumes, whether or not such altered volume of gas actually flowed. This results in price inflators/deflators.

Peak Demand – The maximum amount of gas (electricity) that is consumed during a specified period of time (e.g., an hour).

Pipeline Fuel – Gas consumed in the operation of pipelines, primarily in compressors.

Price Elasticity – The ratio of the percentage change in demand for a good to the percentage change in the price of that good. Demand is considered to be elastic when the ratio exceeds one and inelastic when it is less than one.

Rate Base – The value of a utility's property, established by the IURC, upon which the utility is allowed to earn a specified return.

Rate of Return – The ratio of allowed operating income to a specified rate base expressed as a percentage.

Real – A price that has been adjusted to remove the effects of changes in the purchasing power of the dollar. A real price reflects changes in the value relative to a base year (e.g., 1990).

Real Gross Domestic Product – Real GDP is the figure derived by deflating each component of the GDP for the general level of increase in prices.

Reliability – The assurance of system performance at all times and under all reasonable circumstances to ensure quality, adequacy and economy of gas.

Repressuring – The injection of gas into oil or gas formations to effect greater ultimate recovery.

Residential Consumption – Gas used in private dwellings, including apartments, for heating, cooking, water heating, and other household uses.

Secondary Market – In the gas industry, this is the market for re-selling unneeded pipeline capacity.

SFV Pricing: Straight-Fixed Variable Pricing – The FERC approved SFV for pipelines to have a higher degree of assurance that their costs would be covered by customers. All fixed costs would be allocated to customers according to their peak day entitlement. In other words, SFV rate design allows pipelines to recover most of its costs through the demand component, rather than the commodity component. As a consequence, customers that have a relatively low load factor (peak demand in relation to average use) pay more than those customers that have a relatively constant usage pattern throughout the year.

Shippers – Another name for customers (e.g., industrial, LDC).

SNG: Synthetic Natural Gas – A manufactured product, chemically similar in most respects to natural gas, that results from the conversion or reforming of petroleum hydrocarbons and may easily be substituted for or interchanged with pipeline quality natural gas.

Spot Gas – This is typically gas that is purchased on a short-term basis and is furnished on an "as available" basis.

Storage – Storage may take the form of underground in salt caverns, abandoned gas/oil well or in above ground containment vessels such as liquified natural gas.

Storage Additions – The volume of gas injected or otherwise added to underground natural gas or liquefied natural gas storage during the applicable reporting period.

Storage Withdrawals – Total volume of gas withdrawn from underground storage or liquefied natural gas storage during the applicable reporting period.

Supplemental Gaseous Fuels Supplies – Synthetic natural gas, propane-air, refinery gas, bio-mass gas, air injected for stabilization of heating content, and manufactured gas commingled and distributed with natural gas.

Tcf – Trillion Cubic Feet

Therm – One hundred thousand British thermal units

Unbundling – Generally, this involves the separation of various services upstream of the LDC (e.g., production, transmission and storage). For a LDC, unbundling might include: transportation, metering, billing, storage, backup, and balancing.

Underground Gas Storage Reservoir Capacity – *Inter*state reservoir capacities certified by FERC. Independent producer and *intra*state company reservoir capacities are reported as developed capacity.

Variable Costs – The opposite of fixed costs. These are costs that vary over time (e.g., the cost of purchasing gas).

Vented Gas – Gas released into the air on the base site or at processing plants.

Wet Gas – Unprocessed natural gas or partially processed natural gas, produced from strata containing condensable hydrocarbons and liquid hydrocarbons in solution, and often found in a gathering system prior to processing. This gas has OFTEN not yet been processed. It still may contain water (gas with more than seven pounds of water per Mcf), heavier hydrocarbons (propane, ethane, etc.), and naturally occurring contaminants (sulphur, carbon dioxide), which, mixed with water and/or air, could cause health or physical damage.

Wellhead – It is a term to describe the production fields. The wellhead price of natural gas at the source. Usually, this is the total price delivered to the City Gate minus transportation and storage costs.

Wellhead Price – Represents the wellhead sales price, including charges for natural gas plant liquids subsequently removed from the gas, gathering and compression charges, and State production, severance, or similar charges.

Working (Top Storage) Gas – The volume of gas in an underground storage reservoir above the designed level of the base. It may or may not be completely withdrawn during any particular withdrawal season. Conditions permitting, the total working capacity could be used more than once during any season.

Part 3

A Glossary of General Terms

Aged Fail – Contract that is still not settled between two broker-dealers 30 days after settlement date.

Aggregate Exercise Price – The number of shares in a put or call contract multiplied by the Exercise Price.

Alpha – Coefficient measuring the portion of an investment's return arising from specific (non-market) risk.

Arbitrage – The simultaneous purchase of one commodity against the sale of another in order to profit from fluctuation in the usual price relationships.

At the Market – A futures order placed at the market is executed immediately at the price available when the order reaches the trading floor.

At the Money – An option in which the strike price is nearest the current price of the underlying deliverable.

Backwardation – A market situation in which futures prices are progressively lower in the distant delivery months.

Basis Differential – The difference between the spot market cash price at a certain location and the NYMEX futures price, or the difference between spot cash prices at different locations.

Basis Point – Smallest measure used in quoting yields on bills, notes, and bonds.

Basis Price – Price an investor uses to calculate gains when selling an option.

Bear – One who believes prices will move lower.

Bear Market – A market in which prices are declining.

Beta – A measure of volatility that tells how much a stock moves in relation to an index or average.

Bid and Asked – "Bid" is the highest price a prospective buyer is prepared to pay at a particular time for a trading unit of a given security; "Asked" is the lowest price acceptable to a prospective seller of the same security. Together the two prices constitute a quotation; the difference between the two prices is the spread.

Bid Wanted – Announcement that a holder of securities wants to sell and will entertain bids.

Bull – One who expects prices to rise.

Bull Market – A market in which prices are rising.

Butterfly Spread – Complex option strategy that involves selling two calls and buying two calls on the same or different markets with several maturity dates. One of the options has a higher exercise price and the other has a lower exercise price than the other two options. An investor in a butterfly spread will profit if the underlying security makes no dramatic movements because the premium income will be collected when the options are sold.

Buy Order – In securities trading an order to a broker to purchase a specific quality of a security at the market price or at another stipulated price.

Buyback – Purchase of a long contract to cover a short position, usually arising out of the short sale of a commodity. Also, purchase of identical securities to cover a short sale.

Buyer's Market – A condition of the market in which there is an abundance of goods available and hence buyers can afford to be selective and may be able to buy at less than the price that had previously prevailed.

Call – An option contract that gives the holder the right to buy the underlying security at a specified price for a certain, fixed period of time.

Call Date – Date on which a security may be redeemed before maturity.

Call Option – An option that gives the buyer (holder) the right, but not the obligation, to buy a futures contract (enter into a long futures position) for a specified price within a specified period of time in exchange for a one-time premium payment. It obliges the seller (writer) of an option to sell the underlying futures contract (enter into a short futures position) at the designated price, should the option be exercised at that price.

Call Premium – Amount that the buyer of a call option has to pay to the seller for the right to purchase a stock or stock index at a specified price by a specified date.

Cash Commodity – The actual, physical commodity.

Cash Market – The market for a cash commodity where the actual physical product is traded.

Class of Options – Option contracts of the same type (call or put) and style that cover the same underlying security.

Clear Securities – Comparison of details of a transaction between brokers prior to settlement; final exchange of securities for cash on delivery.

Closing Purchase – A transaction in which the seller's intention is to reduce or eliminate a short position in a given series of options.

Closing Sale – A transaction in which the seller's intention is to reduce or eliminate a long position in a given series of options.

Commitment or Open Interest – The number of futures contracts in existence at any time that have not as yet been satisfied by offsetting sale or purchase or by actual contact delivery.

Contango Market – A market situation in which prices are progressively higher in the succeeding delivery months than in the nearest delivery month.

Contract – Unit of trading for a financial or commodity future. Also, actual bilateral agreement between the parties (buyer and seller) of a futures or options on futures transaction as defined by an exchange.

Cover – To close out a short futures or options position.

Covered Call Option Writing – A strategy in which one sells call options while simultaneously owning an equivalent position in the underlying security.

Covered Put Option Writing – A strategy in which one sells put options and simultaneously is short an equivalent position in the underlying security.

Cycle – A variation where a point of observation returns to its origin.

Covered EFP, or Exchange of Futures for Physicals – A futures contract provision involving the delivery of physical product (which does not necessarily conform to contract specifications in all terms) from one market participant to another and a concomitant assumption of equal and opposite futures positions by the same participants.

Debt Security – Security representing money borrowed that must be repaid and having a fixed amount, a specific maturity or maturities, and usually a specific rate of interest or an original purchase discount. Examples are a bond or a note.

Deep-in-the-Money – A deep-in-the-money call option has the strike price of the option well below the current price of the underlying instrument. A deep-in-the-money put option has the strike price of the option well above the current price of the underlying instrument.

Delta – The amount by which the price of an option changes for every dollar move in the underlying instrument.

Delta Hedging – Hedging method used in option trading and based on the change in premium (option price) caused by a change in the price of the underlying instrument.

Delta Neutral – This is an "options/options" or "options/underlying instrument" position constructed so that it is relatively insensitive to the price movement of the underlying instruments. This is arranged by selecting a calculated ration of offsetting short and long positions.

Derivative Security – A financial security in which the value is derived in part from the value and characteristics of another security, the underlying security.

Equity Options – Options on shares of equity securities.

Exercise – To implement the right under which the holder of an option is entitled to buy (in the case of a call) or sell (in the case of a put) the underlying security.

Exercise Settlement Amount – The difference between the exercise price of the option and the exercise settlement value of the index on the day an exercise notice is tendered, multiplied by the index multiplier.

Expiration – Last day on which an option can be traded.

Expiration Date – The last day in which holders of options must indicate their desire to exercise, if they wish to do so.

Front Month – Trading in the front month expires three to five business days prior to the end of the first calendar day of the delivery month.

Futures – A term used to designate all contracts covering the purchase and sale of financial instruments or physical commodities for future delivery on a commodity futures exchange.

Hedging – A method by which a purchaser or producer of a product uses a derivative position to protect itself against adverse price movements in the cash market.

Historical Volatility – The annualized standard deviation of percent changes in the futures prices over a specified period. It is an indication of past volatility in the marketplace.

Holder – The purchaser of an option.

Implied Volatility – A measurement of the market's expected price range of the underlying commodity futures based on the market-traded option premiums.

In-the-Money – A call option in which the strike price is lower than the stock or futures price, or a put option in which the strike price is higher than the underlying stock or futures price.

Intrinsic Value – The amount by which an option is in-the-money.

Last Trading Day – The final trading day for a particular delivery month futures contract or option contract.

Limit – The maximum amount a futures price may advance or decline in any one day's trading session.

Liquidity – A market is said to "liquid" when it has a high level of trading activity and open interest.

Long Position – The position of a futures contract buyer whose purchase obligates him to accept delivery unless he liquidates his contract with an offsetting sale.

Lot – The standard unit of trading in the futures market.

Near-the Money – An option in which the strike price is close to the current price of the underlying tradable.

Offer – A motion to sell a futures or options contract at a specified price.

Off-peak – Period of low product demand.

Oligopoly – A market with few sellers, but many buyers.

Open Interest – The number of outstanding option contracts in the exchange market or in a particular class or series.

Open Order – A resting order that is good until it is canceled.

Opening Purchase – A transaction in which the purchaser's intention is to create or increase a short position in a given series of options.

Opening Sale – A transaction in which the seller's intention is to create or increase a short position in a given series of options.

Option-Adjusted Spread (OAS) – measures the yield spread of a fixed income instrument that is not attributable to imbedded options.

Out-of-the-Money – A call option in which the exercise price is above the current market price of the underlying security or futures contract.

Over the Counter – Security transactions not performed on a stock exchange.

Parameter – A variable, set of data, or rule that establishes a precise format for a model.

Premium – The price of an option contract, determined in the competitive marketplace, which the buyer of the option pays to the option writer for the rights conveyed by the option contract.

Put – An option contract that gives the holder the right to sell the underlying security at a specified price for a certain, fixed period of time.

Put Option – A contract to sell a specified amount of stock or commodity at an agreed time at the stated exercise price.

Secondary Market – A market that provides for the purchase or sale of previously sold or bought options through closing transactions.

Short Position – A position wherein a person's interest in a particular series of options is as a net writer or seller (i.e., the number of contracts sold exceeds the number of contracts bought).

Range – The difference between the highest and lowest prices recorded during a given trading period.

Roll-over – A special futures straddle trading procedure involving the shift of one month of a straddle into another future month while maintaining the other contract month of the original spread position.

Seller's Market – A condition in the market in which there is a scarcity of goods available and hence sellers can obtain better conditions of sale or higher prices.

Short or Short Position – The market position of a futures contract seller whose sale obligates him to deliver the commodity unless he liquidates his contract by offsetting purchase.

Speculator – An individual who invests in commodity futures with the objective of achieving profits by successfully anticipating price movements.

Spot Market – A market characterized by short-term typically interruptible or best efforts contracts for specified volumes of gas.

Spot Month – The futures contract closest to maturity. The nearby delivery month.

Spread Options – The purchase and sale of two options that vary in terms of type (call or put), strike prices, expiration dates, or both. May also refer to an options contract purchase (sale) and the simultaneous sale (purchase) of a futures contract for the same underlying commodity.

Straddle Options – The purchase or sale of both a put and a call having the same strike price and expiration date. The buyer of a "straddle" benefits from increased volatility, and the seller benefits from decreased volatility.

Stop-Loss – The risk management technique in which the trade is liquidated to halt any further decline in value.

Strike Price – The stated price per share for which the underlying security may be purchased (in the case of a call) or sold (in the case of a put) by the option holder upon exercise of the option contract.

Swap – A custom-tailored, individually negotiated transaction designed to manage financial risk, usually over a period of one to twelve years. Swaps can be conducted directly by two counter-parties, or through a third party such as a bank or brokerage house. The writer of the swap, such as a bank or brokerage house, may elect to assume the risk itself, or manage its own market exposure on an exchange. Swap transactions include interest rate swaps, currency swaps, and price swaps for commodities, including energy and metals. In a typical commodity or price swap, parties exchange payments based on changes in the price of a commodity or a market index, while fixing the price they effectively pay for the physical commodity. The transaction enables each party to manage exposure to commodity prices or index values. Settlements are usually made in cash.

Swing – A set of spreadsheet functions and templates to value the contractual flexibility found in swing, or load-factor, contracts to deliver natural gas or other energy commodities.

Technical Analysis – A form of market analysis that studies demand and supply for securities and commodities based on trading volume and price studies. Using charts and modeling techniques, technicians attempt to identify price trends in a market.

Tick – Refers to change in price, either up or down.

Time Value – The portion of the premium that is attributable to the amount of time remaining until the expiration of the option contract. Time value is whatever value has in addition to its intrinsic value.

Trading Volume – In the futures market, the number of transactions in a contract made during a specified period of time.

Trigger Pricing – A pricing mechanism that allows a party to choose when to "trigger" pricing, which is based on current futures prices at the time of the trigger.

Type – The classification of an option contract as either a put or a call.

Uncovered Call Option Writing – A short call option position in which the writer does not own an equivalent position in the underlying security represented by his/her options contract.

Uncovered Put Option Writing – A short put option position in which the writer does not have a corresponding short position in the underlying security or has not deposited in cash account value of the put.

Underlying Security – The security subject to being purchased or sold upon exercise of the option contract.

Volatility – The market's price range and movement within that range.

Volume – The number of transactions in futures or options on futures contract made during a specified period of time.

Writer – The seller of an option contract.

Part 4

Select Bibliography

Anderson, Dr. John A., Executive Director Electricity Consumers Resource Council *(ELCON)*. Testimony before the Energy and Power Subcommittee, U.S. House of Representatives. (October 1997.)

Audin, Lindsay. "Positioning Your Facility for Utility Competition." *Energy Engineering, Journal of the Association of Energy Engineers,* Dr. Wayne C. Turner, Editor. Oklahoma State University. Stillwater, OK (Vol. 94, No. 2, 1997.)

Beck, R. W. *Form for Request for Power Supply Proposals for City of Hagerstown and Towns of Front Royal, Thurmont and Williamsport, Maryland.* R. W. Beck, Inc. (July 1997.)

Bogart, Jeffrey D. "The Changing World of Utility Investor Relations." *Electrical World.* (January 1996.)

Colucci, Dean M., and Arteaga, Mabel. "How Utilities Can Benefit from Stranded Assets." *Electrical World.* (March 1997.)

Control for the Process Industries; "How to Get Your Project Approved: Know Your Boss." (September 1996.)

Coy, Peter. *Commentary:* "'Stranded Assets': Who Should Foot the Bill?" *Business Week.* (April 28, 1997.)

Department of Energy (DOE), Energy Information Administration (EIA). *Report: Electricity Prices in a Competitive Environment: Marginal Cost Pricing of Generation Services and Financial Status of Electric Utilities.* Reference Case in *Annual Energy Outlook 1997,* DOE/EIA-0383 (97) (Washington, D.C., December 1996.)

Dudley, Susan E. *Using Derivatives to Manage Risk.* Publisher: Edison Electric Institute (EEI). (April 1997.)

Funk, Peter V. K. Jr., and Manley, Michael R. "Deregulation Carries Opportunity and Uncertainty." *Chilton Business News Magazine: Energy User News.* (Vol. 22, No. 8; August 1997.)

Hauck, Gordon. "Buying Electricity – New Deals in Power." Energy User News; *Chilton Business News Magazine.* (Vol. 21, No. 9; September 1996.)

Head, Preston. "Why Use Futures Contracts?" *Electrical World.* (January 1996.)

Hill and Associates, Inc. *Electricity Price Forecast — Outlook for Electricity Market Prices in a Deregulated Environment: 1997-2007.* Publishers: Hill and Associates, Annapolis, Maryland. (October 1997.)

Hughes, John P. "The 10 Myths Against Retail Competition in Electricity," *Iron and Steel Engineer;* (February 1997.)

Hyman, Leonard S., CFA. "Electric Utilities In 2007"; *Hart's Energy Markets – Strategies for Energy Commodities; Power & Profit.* (1997.)

IPALCO, "The Securitization Swindle: High Cost Utilities Use Scheme to Charge Customers for Future Losses." (Excerpts from IPALCO's *Third White Paper.* (May 1997.)

Indiana Utility Regulatory Commission Energy Report (Electricity and Gas) Report No. 1; Section VIII: Emerging Issues in the Natural Gas Industry. (October 1, 1996.)

Inside F.E.R.C.; "Wholesale or Retail, Electric Markets Will Regionalize," The McGraw-Hill Companies, Inc. (September 1997.)

Jones, John. Engage Energy. *"Effective Risk Management Controls Require a Big-Picture View."* (1997)

Kucher, Liane. "Playing for Profits," *Megawatt Markets.* (Fall 1997.)

Macey, Daniel. "Who's Buying Watt?" *Megawatt Markets.* (Summer 1997.)

Nance, Peter K., Teknecon, Inc. *Establishing A Portfolio Management Program.* (1997)

North American Reliability Council Regions (NERC Regional Councils); Aggregated data for NERC regions. (1998)

New York Mercantile Exchange *(NYMEX/COMEX).* Aggregated information from Internet (www.nymex.com.). (December 1997.)

Pilipovic, Dragana. *Energy Risk;* New York: McGraw-Hill, 1998.

Resource Data International, Inc., Boulder, Colorado. *Power Markets in the U. S.* (Vol. 1 and Vol. 2, 1996.)

Rideout, Steve (Integrated Energy Services, LC). "Sophisticated, Unique Technology Revolutionizes Risk Management Industry." *Energy Marketing/A Supplement of PennWell Publishing Co.* (1997)

Tahiliani, Vasu (Enervision), and Rideout, Steve (Integrated Energy Services, LC). "Post-deregulation Risk Management: A New Paradigm for Power Producers and Retail Energy Services Industry." *Energy Marketing/A Supplement of PennWell Publishing Co.* (1997)

Talati, Sunil P. *Transmission System Management & Pricing.* (Draft: August 1997.)

Weiss, Larry. "How Suppliers Can Profit In A Competitive Electric Market," ComPower Group. Aggregated information from Internet. (January 1998.)

Zink, John C., Ph.D.,P.E. "Competition Brings Powerful Changes," *Power Engineering.* (December 1997.)

Part 5

Index